AIR AND SPACE

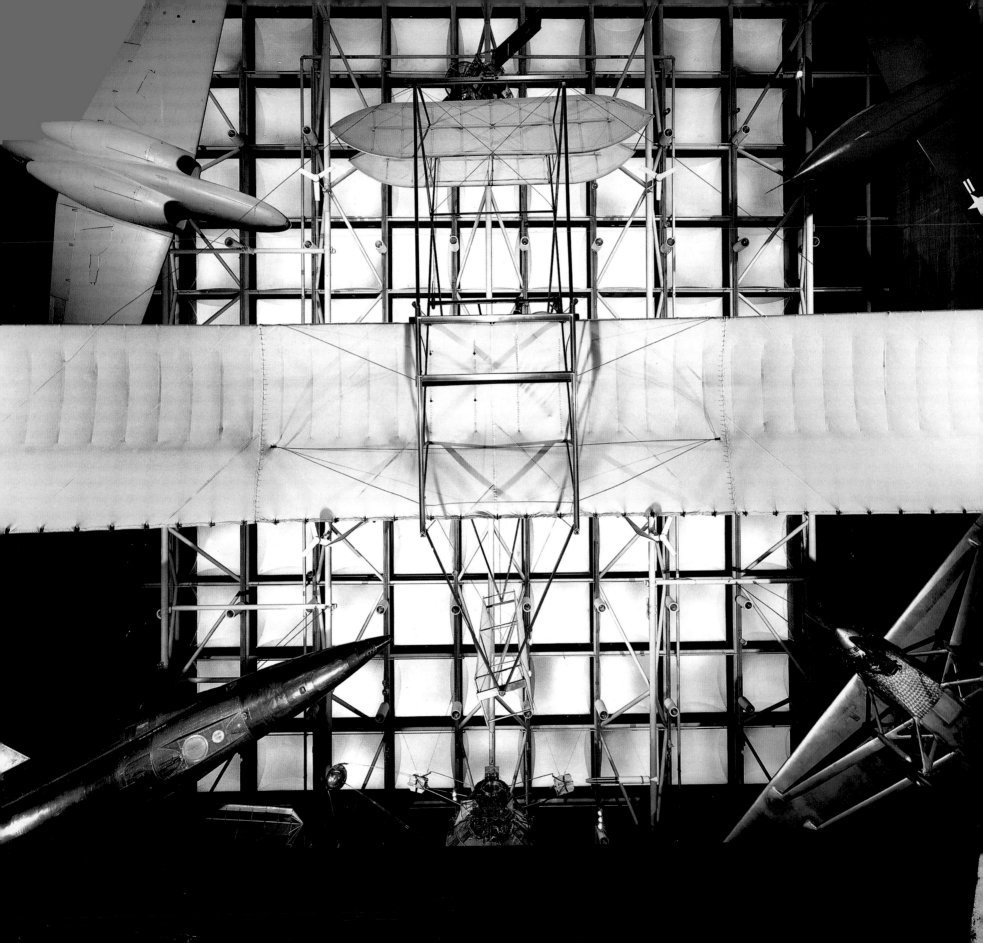

AIR AND SPACE

THE NATIONAL AIR AND SPACE MUSEUM STORY OF FLIGHT

ANDREW CHAIKIN

The National Air and Space Museum, Smithsonian Institution
in association with

BULFINCH PRESS/LITTLE, BROWN AND COMPANY

Boston   New York   Toronto   London

First edition

Text and photo editing by Andrew Chaikin
Captions for Chapters 1–5 were written by Bryan Kennedy

Designed by Janis Owens

LIBRARY OF CONGRESS CATALOGING-IN-PUBLICATION DATA

Chaikin, Andrew.
    Air and space : the National Air and Space Museum's story of flight / Andrew Chaikin — 1st ed.
            p. cm.
    "National Air and Space Museum, Smithsonian Institution, in association with Bulfinch Press."
    Includes index.
    ISBN 0-8212-2082-9
    1. National Air and Space Museum. I. National Air and Space Museum. II. Title.
    TL506.U6W37315   1997
    629.1'074753 — dc20   96-31929

See pages 316–17 for photo credits and acknowledgments

(PAGE II)
The Wright 1903 Flyer (center) dominates this view of the Milestones gallery by Museum photographer Eric Long. The
photograph was taken looking straight up; the ceiling skylights are visible behind the aircraft and artifacts. Arranged around
the Wright Flyer are (clockwise from bottom) a prototype of the Pioneer 10 interplanetary spacecraft; the X-15 rocket plane,
which traveled to the edge of space; the XP-59A Airacomet, the first American jet aircraft; the X-1 rocket plane, which first
broke the sound barrier; and Charles Lindbergh's Spirit of St. Louis.

(OPPOSITE)
The Freedom 7 spacecraft in which Alan Shepard became the first American in space, on May 5, 1961, was on display in the
Apollo to the Moon gallery until it was sent on tour with the "America's Smithsonian" exhibition in 1996.

Bulfinch Press is an imprint and trademark of Little, Brown and Company (Inc.)
Published simultaneously in Canada by Little, Brown & Company (Canada) Limited

PRINTED IN THE UNITED STATES OF AMERICA

CONTENTS

AUTHOR'S PREFACE

When I came to work at the National Air and Space Museum in 1978, a few months out of college, I felt I had arrived at the most enticing workplace in all of Washington. I remember the thrill of entering the Museum's Milestones of Flight gallery before work each day. All around me, icons of aviation and spaceflight glowed in the morning sunlight. Outside, visitors had already begun to gather at the glass doors; soon they would experience the Museum's wonders for themselves.

For a moment, though, I had the Milestones gallery to myself. Alone with the 1903 Wright Flyer, the *Spirit of St. Louis*, the supersonic X-1, and the Apollo 11 command module, I felt I could relive our century's greatest adventure. What began 100 years ago with manned gliders soaring briefly above hillsides has culminated in an age when people cross oceans in a matter of hours, no part of the globe is inaccessible, and human beings have left footprints on the moon.

That feeling was still vivid when I returned to the Museum in the spring of 1995 to tell the story of flight. No writer would be advised to take such an assignment lightly. In the decades since the Wright brothers steered their ingenious creation—the world's first airplane—into a December wind at Kitty Hawk, more has been written about those first flights, and the countless ascents that came afterward, than any single author could absorb in a lifetime.

The focus of this book became clear when I thought of those people who gather each day beyond the glass doors: the vacationers for whom the Air and Space Museum is at the top of the list of things to do in Washington; the staff members from nearby government offices who come through on their lunch hours; the former pilot who flew a plane just like one in the Museum's collection—or, perhaps, the very aircraft that is on display. For all of them, the Museum presents, as no other

(OPPOSITE) The Hall of Air Transportation displays aircraft from the evolution of passenger aviation.

museum in the world can, the extraordinary inventions, innovations, and explorations that have marked the history of flight.

My goal with this book is to create a link in words and pictures between the Museum's collection of aircraft and spacecraft and the history behind them. At best, this has been a daunting task; fortunately, I have been able to rely on the men and women of the National Air and Space Museum, who represent a truly astonishing collection of expertise. They have made the story of flight their life's work, and they have been my greatest resource. It is to them that this book is dedicated.

Andrew Chaikin
Spring 1997

Three boys examine a model of the nuclear-powered aircraft carrier USS *Enterprise* outside the Museum's Sea-Air Operations gallery. Model builder Stephen Henniger spent about 1,000 hours per year for 12 years to create the 1/100 scale model.

AUTHOR'S ACKNOWLEDGMENTS

I am indebted to Don Engen and everyone else at the National Air and Space Museum for their generous assistance with this project. That activity went as smoothly as it did largely because of the patient oversight of Patricia Graboske. Don Lopez, Tom Crouch, and Ted Maxwell also did their share of shepherding the project through many small crises.

Gathering the images for this book was a massive undertaking, and I was fortunate to have the assistance of Melissa Keiser and her colleagues in the Archives Division under the direction of Tom Soapes, especially Brian Nicklas, Kristine Kaske, Allan Janus, Dan Hagedorn, and Dana Bell, as well as Phil Edwards in the SI Libraries. Susan Lawson-Bell was extremely helpful in the selection of artwork. Greg Bryant also lent his expertise to photo research. I appreciate the assistance of Kim Riddle and Walt Ferrell in the Museum's Office of Public Affairs and of Museum photographers Carolyn Russo, Mark Avino, and Eric Long for their efforts in taking and printing photographs for this book.

At Smithsonian Books, Alex Doster, Frances Rowsell, and Laura Kreiss were most generous with their assistance, including the loan of many images used in this book. Special thanks to Bryan Kennedy, whose knowledge of aviation history was invaluable. At NASA, Mike Gentry and his staff were, as always, indispensable. Art Dula generously supplied a number of images of Soviet space activities.

Writing this book has been a journey for me through a marvelous era, and I benefited immeasurably from my guides along the way, the Museum's curatorial staff. In particular, I want to thank Tom Crouch, Tom Alison, John Anderson, Dorothy Cochrane, Ron Davies, Tom Dietz, Von Hardesty, Peter Jakab, Russell Lee, Rick Leyes, Robert van der Linden, Joanne Gernstein London, Michael Neufeld, Dominick Pisano, and Alex Spencer of the Aeronautics Department for many helpful discussions and key ideas. Anita Mason and Collette Williams helped me to feel at home in the Aeronautics Department during many research visits.

Thanks also to Robert Smith, Paul Ceruzzi, Martin Collins, James David, David DeVorkin, Gregg Herken, Cathleen Lewis, Valerie Neal, Allan Needell, and Frank Winter of the Space History Department. Don Lopez and Steven Soter read the manuscript and made many helpful suggestions. I was happy to have the assistance of several of my former colleagues at the Center for Earth and Planetary Studies, including Ted Maxwell, Tom Watters, Pris Strain, and Bob Craddock.

Outside the Museum, I benefited from conversations with a number of historians, in particular, Tom Heppenheimer, whose comments on the development of modern rocketry were especially helpful. Peter Gorin, Asif Siddiqi, Charles Vick, and Michael Cassutt gave me invaluable insight into the history of Soviet human space flights. Museum alumnus Howard Wolko offered his perspective on the technology behind the first modern airliners.

Finally, I am grateful to the participants in the history of flight whom I interviewed—aviators, astronauts, and engineers—for their time and their memories. Special thanks go to Carol Leslie, publisher of Bulfinch Press, Dorothy Williams, my editor, Janis Owens, the designer, and Ann Eiselein, Steve Lamont, Ken Wong, Patricia Hansen, Janice O'Leary, and Mary Reilly; without them this book would not exist.

(LEFT) *Friendship 7*, the Mercury spacecraft in which John Glenn became the first American to orbit Earth, in 1962, now resides in the Milestones of Flight gallery.

(OPPOSITE) Rockets, Skylab, and the Hubble Space Telescope dominate this view of the Museum's Space Hall.

FOREWORD

(ABOVE) Workmen attend to cleaning the 1903 Wright Flyer in the Milestones of Flight gallery.

(OPPOSITE) An aerial view of the National Air and Space Museum in Washington, D.C. The Capitol is visible in the background.

The doors of the National Air and Space Museum have now been open more than twenty years, we have passed the 175 million visitor mark, and our collection continues to expand. A sadder milestone was passed when Paul E. Garber, the museum's "founding father," left us in his ninety-third year (1899–1992). Paul was that person most responsible for the Smithsonian Institution's amazing collection of aviation and spaceflight artifacts, and his legacy will live on for us all to see and enjoy in the future.

The National Air and Space Museum is a museum of technology, history, and research. The developing sciences of aeronautics and space travel have opened great new vistas to man, and the National Air and Space Museum enables visitors to step inside aviation and space history to consider the many facets of that development. Before a Museum exhibit can be brought to life, much must happen. There must be an artifact that is accurately restored in form and quality. While the restoration is done at our Paul E. Garber Facility, the curator does extensive research to ensure a historically correct context for the exhibit. There is always a great deal of discussion within the Museum to ensure an exhibit's authenticity. The curator then works with the Exhibits Department to plan the display of the artifact, and the Exhibits Department designs and constructs the display area. This brief summary describes a complex process that can take three to five years for a major exhibit.

Since President Gerald Ford first opened the doors of our building on July 1, 1976, astronaut Michael Collins, Dr. Noel Hinners, Walter Boyne, Dr. Martin Harwit, and I have been the successive directors. Each of us has been responsible for

XIII

the tone and substance of the Museum during our tenure, and I am greatly pleased to be here.

I feel most fortunate to have flown throughout the past fifty-four years as a naval aviator, an engineering test pilot, civil pilot, aircraft manufacturer, member of the National Transportation Safety Board, and administrator of the Federal Aviation Administration. This background and an abiding interest in the exploration of our own solar system have allowed me to witness the development of aviation and space technology spanning half the history of manned flight. Today when I enter the National Air and Space Museum, I do so with reverence for the technology displayed and for the pioneers who flew these great air- and spacecraft.

The National Air and Space Museum collection is the most extensive of its kind in the world. This could not have been done without the behind-the-scenes work at our Paul E. Garber Facility at Silver Hill, Maryland, where technicians labor to restore each artifact before it is displayed and to maintain it afterward as well. We are proud to be in our third decade of occupancy in our museum building on the Mall and proud that we are developing a history of our own.

Andrew Chaikin has created this book with a view toward linking pictures and artifacts from the Museum's collection with our knowledge of history and of the background of the artifacts. Obviously, Mr. Chaikin cannot cover in its entirety a collection that spans many years and some 1,300 major artifacts as well as countless photographs and works of art, but he has woven historical vignettes around key areas of our collections in an attempt to bring them to life for the reader. These descriptions and background stories add a rich context to the artifacts in our displayed collection, and to this pictorial overview of the Smithsonian Institution's National Air and Space Museum. The reader can have a private tour in its pages as Mr. Chaikin reveals the magnitude of the history of aviation and space technology as collected in one of the greatest museums in the world.

Donald D. Engen, Director
National Air and Space Museum

The *Spirit of St. Louis* hangs in the Milestones of Flight gallery.

Conveying some of the rich history of flight, a Museum docent talks to visitors.

CHAPTER ONE

THE WORK OF BICYCLE MEN

Genius is a word Americans love. This ineffable quality, in a composer, a scientist, or a corporate executive, all but ensures entry into our national pantheon of heroes. But on no one do we bestow it with more reverence than on our inventors. Franklin, Whitney, Bell, Edison—these names conjure up images of the lone pioneer, kissed by inspiration, changing the world. The pantheon received two more heroes on a cold December morning in 1903, on a forlorn North Carolina shore. There, a pair of brothers from Dayton, Ohio, named Wilbur and Orville Wright made history's first sustained powered flights in a craft heavier than air and under the complete control of its pilot. As the inventors of the airplane, the Wright brothers have become icons of American ingenuity. When thinking of these two brothers, the word *genius* easily comes to mind.

And yet somehow the myth persists that these men were just a couple of bicycle mechanics who tinkered together a flying machine in the back room. Nothing could be further from the truth. Wilbur and Orville Wright did possess an inventive genius that set them apart from their rivals, who included one of the nation's most distinguished scientists. But the nature of that genius cannot be stated simply. To understand who these men were, and why they succeeded when so many others failed, is to know the secret of flight. Who were Wilbur and Orville Wright?

Had you met them on a bustling Dayton street about the turn of the century, you wouldn't have thought them remarkable. Conservatively dressed in business suits and bowlers, they were quiet, modest men who did not depend on others for their own advancement. They had been shaped, most of all, within the nurturing and protective circle of their own family. From their mother, who built simple household appliances and toys for her children, they inherited an extraordinary ability to work with machines. From their father, an outspoken and controversial clergyman, Wilbur and Orville learned to set their own course through life and to follow it with unwavering determination. Even after they had come to world prominence, the brothers rarely did anything without advising, and often consulting, their father.

As it happened, neither Wilbur nor Orville had officially graduated from high school. But the power of their intellect, their talent for creative thinking, was apparent in almost anything they did. By the summer of 1888 Orville had established a

WRIGHT FLYER

Restored in 1985, the Wright 1903 Flyer hangs in the place of honor in the Milestones of Flight gallery. When craftsmen stacked the disassembled wing spars in a certain order, the shipping instructions "Wilbur Wright, Elizabeth City, North Carolina" and "Wright Cycle Co., Dayton, Ohio" appeared, written in crayon.

small publishing business in Dayton, featuring a printing press that he had designed and built with help from Wilbur. To say the least, it was unorthodox, constructed out of scrap parts, lengths of firewood, and the frame from a folding buggy roof. But it did the job. One visiting pressman who inspected the machine while it was running remarked, "Well, it works, but I certainly don't see how it does the work."

In 1892 the brothers decided to open a bicycle shop to take advantage of the nationwide craze for "wheeling" that had swept through Dayton. They gained a reputation as skilled mechanics, but Wilbur and Orville weren't content to market existing lines of cycles; in 1895 they began to design and build their own. Here too their talents were evident—in the self-oiling wheel hub they created, in the internal combustion engine they built to power the machine tools in their workshop, and in the quality they put into each of their hand-built cycles. Like their printing business, the Wright Cycle Company enjoyed modest success.

HENSON'S AERIAL CARRIAGE

British inventors William Henson and John Stringfellow patented the idea of a propeller-driven "aerial steam carriage," a logical next step after the train and steamship, as early as 1842. Never built, it appeared widely in prints—here, it is depicted over Hyde Park, London.

(RIGHT) JULES VERNE'S VISION OF FLIGHT

A flying machine depicted by Jules Verne in his 1886 novel *Robur the Conqueror* owed more to imagination than to an understanding of the principles of flight.

(BELOW) TABLEAU DE L'ART AEROSTATIQUE

This detail from a nineteenth-century French poster, "Tableau de l'Art Aerostatique," from the collection of the National Air and Space Museum, celebrates the many colors and fanciful shapes of balloons.

BALLOONS

When Wilbur Wright began to ponder the unknowns of flight in 1899, more than 120 years had passed since human beings first took to the air — not with wings, but in balloons. On October 15, 1783, a Frenchman named J. F. Pilâtre de Rozier stood in a circular gallery at the base of a forty-six-foot-diameter, gaily decorated linen balloon. Straw burned over an iron grate suspended inside the balloon, filling the enclosure with hot air. After preparations were complete, de Rozier rose eighty feet into the sky.

This first manned balloon ascent was the work of the brothers Joseph and Etienne Montgolfier. Earlier, the Montgolfiers' balloon had carried a duck, a rooster, and a sheep, a test flight that indicated the brothers' cautious approach to their work. Now, as de Rozier looked down from his high vantage point, the balloon was tethered to the earth for conservatism's sake. Finally, on November 21, de Rozier was joined by François Laurent, marquis d'Arlandes, for the first untethered balloon flight. The pair drifted over Paris for a distance of about five miles, reaching heights of perhaps 300 feet, before returning to Earth. This brief voyage, simple though it was, opened the age of flight.

The surprising thing is that, according to the evidence, no such flight had taken place long before. Anyone who stares into a fire and observes embers lofted above the flames can see that hot air rises. Presumably, the ancients noticed this but did not comprehend what they saw, or how to harness it.

By the late 1700s that knowledge existed. The sciences of physics and analytical chemistry had revealed that air was a mixture of gases, some of which were lighter than others. Theorists had speculated that these differences could be exploited: filling a balloon with light gas makes it less dense than the air around it, causing it to rise. The Montgolfier brothers thought that by burning straw they were creating one such gas; they did not understand that simply heating air decreases its density.

More astute was physicist Jacques Charles, who had learned of the existence of "inflammable air," the gas we

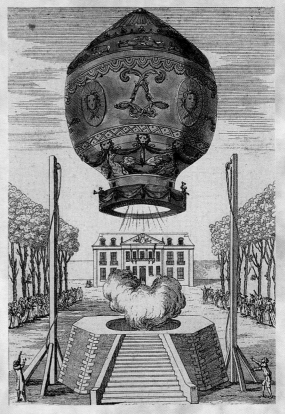

MONTGOLFIER BALLOON, FRANCE, 1783

What the Wrights were to the airplane, papermakers Joseph and Etienne Montgolfier were to the hot-air balloon. This lithograph depicts the November 1783 flight of volunteers Pilâtre de Rozier and François Laurent, marquis d'Arlandes, in a Montgolfier balloon from the Château de la Muette in the Bois de Boulogne to the present-day Place d'Italie, Paris. At left: a model of the Montgolfier balloon is on display at the Museum.

(RIGHT) FIRST AMERICAN PHOTO OF BALLOON

Nineteenth-century chemistry not only provided balloonists with the hydrogen gas needed to fill their balloons, it also made possible the marvel of photography. Here, John Steiner's balloon is inflated in Erie, Pennsylvania. This 1857 ambrotype is the oldest known photograph of American ballooning.

(LEFT) CIVIL WAR BALLOON *INTREPID*

Union Army handlers maneuver the observation balloon *Intrepid* during the Civil War's 1862 Peninsular campaign. Smithsonian secretary Joseph Henry introduced President Lincoln to balloonist Thaddeus S. C. Lowe—which led to the Federal Army's deployment of balloons and created a challenging target for Confederate artillery gunners.

know as hydrogen. Even as the Montgolfiers made their initial experiments, Charles was testing his own hydrogen-filled balloon, made of silk coated with a mixture of rubber gum and linseed oil. Among those who witnessed an unmanned flight in August 1783 was Benjamin Franklin. It was then that Franklin, queried on the usefulness of the new invention, uttered his famous reply, "Of what use is a newborn babe?"

In December 1783, just ten days after the Montgolfier balloon's first free flight, Charles and a companion rode his hydrogen balloon from Paris to Nesle, twenty-seven miles away. To control their ascent and descent, the men used ballast, along with a vent for releasing some of the gas. Remarkably, two technologies for ballooning had emerged simultaneously.

Within a few years, enthusiasm for the new inventions spread throughout Europe and to America. Balloons captured the public's imagination and created something of a craze. By the first years of the nineteenth century, balloons had found roles as observation posts in warfare and as vehicles for scientific exploration of the atmosphere.

The basic balloon technologies that were in place by the end of 1783 are the same ones used today, although refinements have been made in the intervening centuries. In hot-air balloons, the propane tank has replaced burning straw; gas balloons now use helium instead of hydrogen. But the legacy of the wondrous year 1783 still holds a place in aviation and in the imagination of those who witness balloon flights.

If the Wright brothers had been content to continue as printers and bicycle manufacturers, the invention of the airplane would have been left to someone else. But Wilbur Wright was not content; he harbored a growing unease over his situation in life. He disparaged his own abilities as a businessman; he felt he had never found his calling. Wilbur's education had been interrupted by injury and illness and by the need to care for his dying mother, who succumbed to tuberculosis in 1889. He craved a new challenge, one that would allow him to make full use of his talents and abilities. And when he began to find it, in the spring of 1899, it was not in a new business venture, but in the age-old dream of human flight.

It was Bishop Wright who had sparked his young sons' fascination with flight, with a toy—a simple, rubber band–powered helicopter—that he brought back from one of his church trips in 1878. This interest took on new importance in 1896, when Wilbur read of the death of Otto Lilienthal, a German aerial experimenter. A true aeronautical engineer, Lilienthal had delved into the theory of flight and applied what he had learned to gliders that he designed, built, and flew in Germany. His glides astounded the few journalists and spectators who witnessed, as one wrote, "the wild and fearless rush of Otto Lilienthal through the air." But one Sunday in the summer of 1896, Lilienthal's glider suffered a gust of wind and nosed up sharply, then fell to the ground from a height of fifty feet. The crash broke Lilienthal's spine; he died the next day in his hospital bed. Wilbur, who had followed Lilienthal's progress with great interest, noted the loss of the man he would later call "the greatest of the precursors." Lilienthal's death planted a seed of curiosity in Wilbur that flowered in the spring of 1899, when he was looking through a book on ornithology. Nothing in the book revealed how birds, or any other creature, managed to

(ABOVE) **ADER'S *ÉOLE***

French electrical engineer Clément Ader constructed the bat-like *Éole* in 1890. According to some reports, it wobbled 165 feet through the air with Ader as passenger. His later experiments were less successful.

(BELOW) **MAXIM'S FLYING MACHINE**

Machine gun inventor Sir Hiram Maxim dabbled in aviation on a massive scale. His four-ton, twin-steam-engine rig had a crew of three and flew briefly in 1894. Ironically, it was attached to a track that prevented it from flying freely.

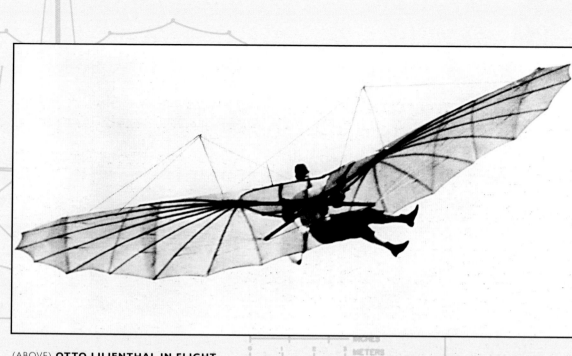

(ABOVE) OTTO LILIENTHAL IN FLIGHT

German mechanical engineer Otto Lilienthal shifts his weight to maneuver his *normal-sagelapparat* (standard sailing machine). The first to methodically study aeronautics, Lilienthal made more than 2,500 glides and published the first data on aerodynamic lift—used by the Wrights and others.

(BELOW) LILIENTHAL STANDARD GLIDER AT THE MUSEUM

Built in 1894, this Lilienthal Standard Glider was purchased by William Randolph Hearst and used for publicity flights until Lilienthal's 1896 death in a similar machine. Constructed of willow and bamboo, it was restored in 1977 and currently hangs in the National Air and Space Museum.

WILBUR AND ORVILLE WRIGHT, 1909
More than three years separated the brothers from Dayton, Ohio. Their sober Midwestern appearance belies their ingenuity in tackling the problems of powered flight. Wilbur (left) and Orville's mathematician mother encouraged their research, and their powers of observation and mechanical abilities were unrivaled.

take to the air. And yet, he reasoned, if countless species as dissimilar as insects, bats, and birds knew the secret of flight, then couldn't human beings learn it as well?

On the face of it, Wilbur's quest might have seemed misguided. The newspapers had been filled with stories of crackpot inventors pursuing the same dream. But Wilbur Wright didn't enter into things lightly; he would pursue this goal with characteristic planning and thought. First of all, he needed information, and he knew where to find it. The secretary of the Smithsonian Institution, Samuel Pierpont Langley, was engaged in his own ongoing flight experiments. Already he had directed the construction of several steam-powered winged models, two of which had made test flights over the Potomac. On stationery from the Wright Cycle Company, Wilbur drafted a letter to the Smithsonian asking for guidance in the problem of human flight. "I wish to avail myself," he wrote, "of all that is already known and then if possible add my mite to help on the future worker who will attain final success."

The Smithsonian responded with a collection of pamphlets on recent research and recommendations for further reading. Within three months Wilbur had digested all of it, the foundations of nineteenth-century aeronautical theory. He now understood that a successful flying machine

must have three basic elements, as first described by nineteenth-century Englishman George Cayley: there must be a structure to produce lift, a method of propulsion, and some means of controlling the craft's flight. By 1899 two of these appeared to be solved. Lilienthal's research had validated the long-held belief that cambered, or curved, wings were the best lift-producing structures. And Langley's miniatures showed that powered flight was truly possible.

Wilbur found that, remarkably, the issue of control had been largely neglected. To steer his sometimes unruly gliders,

WILBUR WRIGHT'S LETTER TO THE SMITHSONIAN

"I am an enthusiast, but not a crank," wrote Wilbur Wright to the Smithsonian Institution in 1899 in search of aeronautical literature. Intuitive engineers, the Wrights knew that controlling an aircraft was important to making it fly successfully.

Lilienthal had employed only the acrobatic motions of his own body. Langley's powered models had been incapable of maneuvering. No one seemed to realize that all three of Cayley's attributes had to be mastered for a working airplane. Although Wilbur Wright had never been trained as an engineer, he thought like one. By

(ABOVE) LANGLEY AERODROME IN WORKSHOP

Langley's model, with its unique steam power plant, was photographed in a workshop behind the Smithsonian building. Langley would devote six years to developing a "Great Aerodrome," unaware of its aerodynamic flaws. The Aerodrome was unsteerable.

(RIGHT) SAMUEL PIERPONT LANGLEY

Third secretary of the Smithsonian Samuel Pierpont Langley built the highly successful, steam-powered Aerodrome No. 5, which flew for half a mile on the Potomac River in 1896. Impressed by his tests, Secretary of the Navy Theodore Roosevelt secured Langley $50,000 to build a manned Aerodrome.

12

Aerodrome "A" Plan.

LANGLEY AERODROME NO. 5

Restored at the Air and Space Museum's Paul E. Garber facility, Aerodrome No. 5 hangs in the Museum. The craft's twin pusher propellers and wing dihedral are clearly visible in this photograph.

LANGLEY AERODROME IN FLIGHT (1895)

As depicted in a painting from the Museum's collection by artist Garnet W. Jex, Samuel Langley's steam-powered model Aerodrome No. 5 takes flight over the Potomac. The news that an object this large had flown stunned the nineteenth-century public.

identifying control as the single most important problem, he focused his effort and avoided getting sidetracked by the technical details that had lured so many others. Now came the challenge of solving it.

Most experimenters assumed that an airplane must be inherently stable. A human pilot, they believed, wouldn't be capable of responding fast enough to sudden gusts of wind. Langley, for example, had designed his aerodromes to automatically right themselves. Wilbur saw the problem differently. He could not have known of the words of a Binghamton, New York, newspaper editor who wrote in 1896 that the invention of the airplane would be the work of bicycle men. Still, he intuitively understood that an airplane, like a bicycle, need not be inherently stable to be controllable. Any newcomer to the sport knows how unstable a bicycle can be, but after some practice the rider learns to control his or her path with precision. So it would be with flying airplanes, Wilbur reasoned. Ideally, the pilot would have a means of controlling the craft in all three axes of rotation, namely roll, pitch, and yaw.

But how to achieve this? Lilienthal's method of shifting his weight was unacceptable. For one thing, it limited the size of the craft and thus the amount of lift the wings could provide. And for Lilienthal it had proved deadly. Control, Wilbur knew, had to be achieved by aerodynamic means.

Pitch and yaw could surely be handled by movable rudders, but what about roll? One day, working alone in the bicycle shop, Wilbur absentmindedly twisted the ends of a long cardboard inner-tube box, and suddenly he had the answer. Wilbur could visualize the solution in his mind. It had come from the birds themselves. He'd seen buzzards in flight change their orientation, to make a turn or to right themselves in a gust of wind, by bending the tips of their wings. Now he realized that if the entire structure of a glider's wing could be warped, or twisted—like the cardboard box—the same effect could be achieved. When he built a kite with a five-foot span to test his wing-warping idea, it worked.

Wilbur's excitement over this success easily spread to Orville; together, the pair decided to collaborate on a man-carrying version of Wilbur's kite. Among their advantages was the close working relationship they had forged over the past decade. Now they would build on the foundation laid by previous experimenters. As a starting point, the brothers chose the design of a biplane glider built in 1896 by French-born Octave Chanute, largely because it would easily accommodate Wilbur's wing-warping concept. In their mathematical calculations, the Wrights made extensive use of Lilienthal's experimental data on the amount of lift and drag generated by an

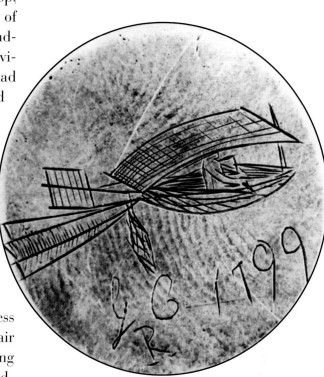

CAYLEY'S GLIDER

Sir George Cayley engraved his strikingly modern concept of an aircraft on this silver disk in 1799. Cayley studied the concepts of lift, thrust, and drag; noted the importance of active control and streamlining; and even understood the shift in the center of pressure as a wing's angle of attack increased.

(LEFT) CHANUTE'S GLIDER

Master engineer Octave Chanute combined his knowledge of bridge building with his passion for aviation. His 1897 biplane glider, created with the help of protégé Augustus Herring, shown here on the shore of Lake Michigan, employed a Pratt truss wing structure. This design would influence aircraft designers for years to come.

(ABOVE) RESCUE CREW

Unlikely spectators, the hardy Kitty Hawk lifesaving crew was photographed in 1900. Three years later, a group of their comrades would witness the first successful flight of a powered, heavier-than-air craft.

airfoil at varying angles to the wind. But in one important way, the Wrights' glider was something new. The wings of Lilienthal's craft, and Chanute's, had a cross section shaped like the arc of a circle, with the highest point in the middle. Working on intuition, Wilbur and Orville shifted the peak of the arch to a point near the wing's leading edge. And they changed the wing's camber, or depth of curvature, to be roughly half that of Lilienthal's and other

gliders. Here was another key to the brothers' success: they were forging their own path. In August 1900, with the peak bicycle sales season over, the new glider was nearly finished.

Leaving Dayton in September with parts of the unassembled glider, Wilbur journeyed to Kitty Hawk on North Carolina's Outer Banks. Orville soon followed. They had chosen Kitty Hawk not for its remoteness, though its isolation would ensure privacy, but for its strong and frequent winds. Together, the brothers finished assembling the glider and in October tested it as a tethered kite, gathering data on its performance. Occasionally, one of them rode the craft for a few minutes to try the controls. And on October 20, after he and Orville had carried the glider up one of three giant dunes known as Kill Devil Hills, Wilbur flew the glider. With characteristic caution, he began with brief, low glides, working up to ones lasting between fifteen and twenty seconds. That was enough to confirm that the wing-warping system, which Wilbur activated by nudging a lever with his feet, effectively controlled roll. Another innovation, the so-called forward elevator, gave him deft control of pitch simply by moving a lever—a far cry from the acrobatics Lilienthal had needed. Ultimately, though, the glider was a disappointment: its wings generated far less lift than the Wrights had calculated. Before leaving Kitty Hawk on October 23, Wilbur and Orville made plans to come back the following summer with a new glider.

July 1901 saw the Wrights' return to Kitty Hawk—leaving behind newly hired assistant Charlie Taylor to mind the bicycle store—with a glider larger and heavier than any previously flown. As they constructed a shed for the new craft, their anticipation was dampened only by swarms of mosquitoes. ("They chewed us

WRIGHT 1900 GLIDER

Flying as a kite in this picture, the Wright brothers' 1900 glider showed promise. It had a seventeen-foot wingspan, Chanute-type wing layout, and front elevator to protect the pilot in the event of a crash.

clear through our underwear and socks," wrote Orville to his sister, Katherine.) On July 27 Wilbur was in the air once again, and there were problems right away. Instead of gliding smoothly, the new machine undulated, sometimes wildly, with Wilbur frantically working the forward elevator and shifting his body weight to avoid crashing. Searching for a culprit, the brothers suspected the new wing design. Fortunately, Wilbur and Orville, with characteristic foresight, had designed this glider so that they could reconfigure the wings to a flatter camber. They did so—

and the craft was immediately responsive. In glides as long as 389 feet, Wilbur easily maneuvered to trace the undulations of the dunes. But the wings produced no more lift than those of the 1900 glider, despite having 75 percent more area. How could their calculations have been so wrong? As the 1901 trials progressed, just as discouraging were complications that emerged with the wing-warping mechanism. Sometimes the glider stalled and turned violently, catching a wingtip in the sand. The broth-ers were a dejected pair as they left Kitty Hawk in August, earlier than planned. On the train home Wilbur declared, "Not within a thousand years will man ever fly."

Wilbur's despair didn't last; both he and his brother were possessed with a single-mindedness that sustained them in the face of adversity. Back in Dayton, Wilbur and Orville pondered the cause of their set-backs and focused on Lilienthal's aeronau-tical data, which they had used in designing their gliders. Could those data

(ABOVE) **WRIGHT 1901 GLIDER**

The ill-mannered 1901 glider was photographed at Kill Devil Hills, Kitty Hawk, North Carolina, when Octave Chanute was watching. Its performance was disappointing. Wilbur and Orville realized that Lilienthal's lift data might be wrong.

(OPPOSITE) **WILBUR FLYING 1902 GLIDER**

This unique photograph shows Wilbur banking into a turn in the 1902 glider at Kill Devil Hills. In addition to its improved wing design, the glider featured a new fixed vertical tail designed to help in the turns. Only after it was hinged, however, and linked to the wing-warping mecha-nism was the tail effective.

(ABOVE) **REPLICA OF THE WRIGHT WIND TUNNEL**

With their homemade wind tunnel, the Wright brothers conducted their historic research into aerodynamics. No photographs of the original wind tunnel exist; this replica is one of a handful that were constructed in the 1950s. It is on display in the Museum's Early Flight gallery.

(LEFT) **WRIGHT WIND TUNNEL BALANCE**

Makeshift in appearance, this precise instrument helped the Wrights isolate the various aerodynamic forces acting on their airfoil models. Essentially, the lift balance (left) and the drag balance were mechanical analogs for lift and drag equations.

be in error? Lilienthal's authority had been unquestioned. Then too, the Wrights had chosen a different shape for their wing than Lilienthal's. The only way to resolve the issue was to conduct their own experiments. Initially, they constructed a simple but elegant device, which they mounted on a bicycle. From these initial tests Wilbur and Orville felt that they could not rely on Lilienthal's data; they must gather their own.

The Wrights realized that they need not construct full-scale gliders for their tests; they could obtain valid results using miniature airfoils. Soon they built a small, homemade wind tunnel—a little-used device whose potential, up to now, had never been fully exploited—powered by the small gasoline engine used to run the bicycle-shop machinery. In the back room of the bicycle shop, Wilbur and Orville spent the fall of 1901 crossing a frontier. Unlike previous experimenters, the Wrights used their wind tunnel not for pure research but to design an airplane, paving the way for generations of aeronautical engineers. As much as their experiments at Kitty Hawk, these tests revealed the brothers' genius for grappling with the complex unknowns of flight. To evaluate their model airfoils, the Wrights built delicate devices, called balances, which served as physical representations of the mathematical relationships described in the equa-

tions of lift and drag. By year's end they had gathered data on hundreds of different airfoils, a trove of the most accurate aeronautical data yet gathered. It was every bit as exhilarating as flying the gliders. "Wilbur and I," Orville would remember, "could hardly wait for morning to come, to get at something that interested us. *That's* happiness!"

But these brothers were never interested in research for its own sake. Always their eyes were on their goal: a successful flying machine. No sooner had they finished their wind-tunnel experiments than they were incorporating their findings into their third manned glider. In September 1902 they assembled the craft at Kitty Hawk and found that it performed even better than expected, with one exception. The brothers had designed a new vertical rudder to prevent the previous summer's turning difficulties. Instead, it seemed to make things worse. Already the two experimenters had traced the problems with turns to the increased drag produced by wing-warping. Now Orville—who this summer had finally begun to fly—hit on a solution: a hinged rudder that could be used to counteract the extra drag. Wilbur's innovation was to link the rudder's controls with the one for the wing-warping, so that they worked together.

The new rudder did the trick. By the end of October, Wilbur and Orville had

made hundreds of glides, including a spectacular, record-setting one of more than 622 feet in twenty-six seconds. They controlled their craft in all three axes, maneuvering with ease and precision. The Wrights could now see, as no inventor before them had, that their goal was within reach. They had solved the basic problems of flight; all that remained was to add a means of propulsion. By the time they left Kitty Hawk, they had already calculated the basic design for the world's first airplane.

Meanwhile, the Wrights' most prestigious competitor, Samuel Langley, was gearing up for his own experiments in manned powered flight. The Great Aerodrome, as Langley's flying machine was called, was nearing completion after an effort spanning almost two decades and costing more than $50,000. As secretary of the Smithsonian, Langley was America's chief scientist, with the nation's entire scientific establishment at his disposal. But his approach to the problem of powered flight was seriously flawed. The Aerodrome was so flimsy that simply pressing on the frame would bend it out of shape. Its minimal control system had never been tested. Langley was obsessed with propulsion, and much of his effort and funds had gone to developing a powerful gasoline engine.

In contrast, the two bicycle men from

Dayton had crafted such an efficient flying machine, they calculated that a gasoline engine producing a minimum of only eight horsepower was needed for their propulsion system. They built it themselves, with help from Charlie Taylor, when there proved to be no manufacturer interested in the job at a reasonable price. Aside from the crankcase, which was unusual because it was made of cast aluminum, the Wrights used no exotic materials for their flying machine. As with the gliders, its frame was spruce and ash covered by muslin; the crankshafts were steel tube. What made the difference was the thinking behind it.

In the spring of 1903 the Wrights put their creative abilities to the test once again, this time on the matter of a propeller. Previous experimenters hadn't expended much effort on this issue, opting for simple angled blades. Initially, Wilbur and Orville planned to adapt whatever design theories existed for a ship's propeller, which is essentially a screw advancing through water. Finding that there were no such theories—the design of ships' propellers was an empirical business—the brothers attacked the problem from scratch. Their analysis yielded the crucial insight that an airplane propeller would be something very different, a rotating wing whose lift moves the plane forward. Designing such a propeller was an astonishingly complex task. A host of variables,

including the speed of the blades' rotation, their angle of attack, the forward speed of the airplane, and wind speed, were interrelated. The more the Wrights delved into the puzzle, the more intractable it seemed. The bicycle shop was the scene of some vigorous and often loud disagreements. At times the brothers argued their positions so passionately that each man was converted to the other's point of view. The disputes startled Charlie Taylor, but they were crucial to uncovering solutions. And to the brothers themselves, they were invigorating. "I love to scrap with Orv," Wilbur once said. "Orv is such a good scrapper."

By September the machine they called "the Flyer" was finished, and the brothers prepared to leave for Kitty Hawk, not merely as aerial experimenters but as the world's leading aeronautical researchers. Orville would later write to a friend, "Isn't it astonishing that all of these secrets have been preserved for so many years just so that we could discover them!!"

Meanwhile, in Virginia Langley's Aerodrome was nearing its first test flight. It would be launched by a massive catapult system from atop a houseboat in the Potomac River. Langley's assistant, Charles Manly, who had played a key role in developing the Aerodrome's engine, would be the pilot; in truth, he would be little more than a passenger. Suspended in a fabric

LANGLEY AND MANLY

America's leading scientist was photographed on the houseboat that served as launch platform for the Aerodrome; he stands next to brave pilot Charles Manly. Manly played an important role in the design of an efficient radial engine for the large Aerodrome—its only redeeming feature. Little did Manly know the disaster that awaited him.

enclosure on the craft's underside, Manly would end up submerged in the Potomac even if the flight were successful.

The Aerodrome's first attempt to fly, on October 7, 1903, was over almost as quickly as it began. With Manly perched in the tiny cockpit, and with the engine running at full speed, the craft was flung forward. The Aerodrome plunged into the water, in the words of one witness, "like a handful of mortar." Manly and Langley (who was not present) both blamed a flaw in the launching system and turned their energies to repairing the Aerodrome in time for a second attempt before year's end.

The Wright brothers were at Kitty Hawk when they learned of Langley's attempt. "I see that Langley has had his fling," Wilbur wrote Octave Chanute. "It seems to be our turn to throw now, and I wonder what our luck will be." So far, their luck had been poor. Preparations had been hampered by relentless storms and by a balky assistant (local Dan Tate, whom the Wrights had hired at the princely sum of seven dollars per week). The engine, tested in early November, ran so rough that it damaged the propeller shafts, which had to be sent back to Dayton for repair. The engine ran smoother at month's end, but one of the repaired shafts cracked during a test run. This time Orville journeyed home to create a new set out of spring steel.

(ABOVE) **LANGLEY AERODROME ON LAUNCH PLATFORM ATOP HOUSEBOAT**

Contemporary photograph shows the floating launcher and giant Aerodrome. Initially, the failure of Langley's ship would be blamed on the elaborate launch mechanism. In fact, it was simply too weak structurally to hold together.

(RIGHT) **AERODROME FAILURE**

The *Washington Post* gleefully reports the crash of the Aerodrome. On its second flight attempt, the rear wings buckled and Manly nearly drowned in the frigid Potomac. Langley was humiliated by the failure.

COLLAPSE OF THE AIRSHIP

AIRSHIP FAILS TO FLY

Prof. Laugley's Machine Goes to River Bottom.

PROF. MANLY ABOARD

THE LATTER RESCUED FROM PERILOUS POSITION.

23

Orville finished his work on December 8, the same day that Langley's Aerodrome was poised for one more try outside Washington, D.C. While Langley watched, the Aerodrome sped down its track and, in a sickening instant, flipped over on its back and fell into the river. Manly was pulled from the water uninjured but so chilled that his clothing had to be cut from his body. The real casualty that day was Samuel Langley. His ambition to create a flying machine had crumbled. For the rest of his life, he would endure ridicule for his failure.

Orville Wright arrived back at camp on December 11. On the fourteenth the winds were strong enough to attempt a flight. The machine accelerated down its launch rail and suddenly leaped upward, only to stall and crash to the sand, breaking one of the supports for the forward elevator. Wilbur, who had won a coin toss and was at the controls, was unhurt. He and Orville had learned the hard way that the craft's pitch control was extremely sensitive. And they also knew what lay ahead. The next day the brothers sent their family a brief telegram ending with the words

SUCCESS ASSURED KEEP QUIET.

By December 17 the machine was repaired and the winds had picked up. The temperature had plummeted to the freezing mark. With the Atlantic winter approaching, the Wrights could not afford to wait any longer. At 10:30 A.M., in a twenty-seven-mile-per-hour wind and with several men from the nearby lifesaving station as witnesses, Wilbur and Orville started the Flyer's engine, then backed off to let it warm up. It was Orville's turn to pilot the craft, but just now he and Wilbur exchanged words and shook hands. One witness noticed, "They held on to each other's hand, sort o' like two folks parting who weren't sure they'd ever see one another again."

Finally, Orville took his place, lying prone amid the staccato din of the engine. Wilbur, stationed at the end of one wing, was prepared to steady the craft as it moved down the track. At 10:35 the Flyer began to move, picking up speed, and at last rose into the air. While Wilbur watched in amazement, it sailed jerkily above the sand, then came to rest again. The machine had traveled 120 feet in about twelve seconds. The brothers had to estimate the time because the Flyer's onboard timer had been reset by the impact of landing, and because Wilbur Wright—the most exacting of engineers—had been so excited that he neglected to start his own stopwatch.

Before the morning was over, the brothers had made three more flights, the longest lasting nearly a minute and covering 852 feet. The airplane was a reality. Journeying to the lifesaving station, they telegraphed home:

SUCCESS FOUR FLIGHTS...

The text went on to describe the average speed and the duration of the longest flight. It ended,

INFORM PRESS HOME CHRISTMAS.

The last request proved to be more difficult even than flying had been. The editor of the Dayton *Journal* quipped, "Fifty-seven seconds, hey? If it had been fifty-seven minutes, then it might have been a news item."

An astonishing reaction, and yet it was not unusual. After the much-publicized failures of Langley and others, many who read the exaggerated newspaper accounts of the flights didn't believe them. In the spring of 1904 a few reporters witnessed Orville making a twenty-five-foot hop with a new Flyer at a cow pasture outside Dayton but came away relatively unimpressed. What little news emerged of the brothers' continuing experiments could not compete with reports from France, where an aviator named Santos-Dumont steered his airship on lengthy flights over Paris. Even as the Wrights went on to perfect their invention with the 1905 Flyer—the first fully

(OPPOSITE) **WRIGHT FLYER'S FIRST FLIGHT, 1903**
Elevator pitched up for altitude, the world's first successful heavier-than-air aircraft was captured on film at the moment of takeoff. The Wright 1903 Flyer rode down the wooden launch rail on a dolly, which can be seen at the end of this rail, then took off into the air.

(LEFT) **WRIGHT 1903 FLYER AT THE MUSEUM**

A close-up of the Orville Wright mannequin on the restored Flyer at the Museum. His hand rests on the elevator control lever, and the flight-data recording instruments are mounted on the strut to his right.

(BELOW) **WILBUR IN 1901 GLIDER**

"Not within a thousand years will man ever fly!" cried Wilbur Wright, over the erratic behavior of their 1901 glider during flight tests at Kitty Hawk. Lacking the expected lift, it also tended to slide dangerously sideways into the ground during turns.

THE WESTERN UNION TELEGRAPH COMPANY.
INCORPORATED
23,000 OFFICES IN AMERICA.   CABLE SERVICE TO ALL THE WORLD.

RECEIVED at

176 C KA CS 33 Paid.   Via Norfolk Va

Kitty Hawk N C Dec 17

Bishop M Wright
                    7 Hawthorne St

Success four flights thursday morning all against twenty one mile
wind started from Level with engine power alone average speed
through air thirty one miles longest 57 seconds inform Press
home Christmas .
                              Orevelle Wright   525P

SUCCESS AT LAST

The Wright brothers' telegram home to their father in Dayton was almost anticlimactic. The press, for its part, refused to believe the gentlemen from Ohio or simply ignored them. How could bicycle mechanics succeed where Langley had failed?

controllable airplane—they received little recognition in their own country. Not until 1908, when Wilbur journeyed to France, did the world believe what had happened at Kitty Hawk five years earlier. And the most difficult struggles lay ahead, including a battle to claim patents for their flying machine. The secrecy the Wrights enforced regarding their invention only delayed its recognition. Ultimately, years of legal disputes so weakened Wilbur Wright that he succumbed to typhoid fever in 1912. Orville, carrying on alone, later fought a feud with the Smithsonian, which insisted that Langley had been the first experimenter "capable of flight." The dispute was not resolved until 1943. In 1948, the year of Orville's death, the Flyer was un-veiled at the National Museum. The exhibit label—which still accompanies the Wright Flyer where it is displayed today, in the National Air and Space Museum's Milestones of Flight gallery—left no doubts that the feud had been resolved. And it came close to stating, in one paragraph, the genius of Wilbur and Orville Wright: "By original scientific research the Wright Brothers discovered the principles of human flight…. As inventors, builders, and fliers they further developed the aeroplane, taught man to fly, and opened the era of aviation." What took place at Kitty Hawk on that chilly December morning was not merely a technological quantum leap but a turning point in human evolution. Human beings, who for millennia had watched the birds with envy, were now fliers. ■

(LEFT) **1903 FLYER ON ITS WAY HOME IN U.S. NAVY TRAILER, 1948**

Amazingly, the Wright 1903 Flyer spent the years 1928–48 in Europe. The Smithsonian's refusal to acknowledge their achievement led surviving Wright brother Orville to send this American icon to England.

(RIGHT) **PAUL GARBER IN 1903 FLYER**

Smithsonian employee Paul Garber tries out the Flyer he helped recover after World War II. More than any other person, Garber worked to heal the breach between the Museum and the surviving father of heavier-than-air flight.

WRIGHT 1905 FLYER IN FLIGHT

The Wright 1905 Flyer, the first practical airplane, is shown during one of its many record-setting flights. On October 4, 1905, it stayed aloft for 33 minutes and 17 seconds and covered twenty miles, an achievement that was not fully appreciated until years later.

CHAPTER TWO

ALL THINGS POSSIBLE

The French people were in for a shock in August 1908. For almost two years, ever since a Brazilian inventor had flown an ungainly contraption in a Paris park, they had been secure in the belief that France was the birthplace of aviation. But in August 1908 Wilbur Wright went to France, and before he left, he would set his many doubters straight. His triumph there, arguably the true beginning of aviation's first decade, began an era of such unbridled progress that there seemed to be no limit to the airplane's potential to change the world.

Before 1908, with the Wrights' penchant for secrecy, only the sketchiest details of their Flyer and its flights had emerged; these were not enough to convince the French. Late in 1905 there were reports that the American brothers had flown thirty-nine kilometers (twenty-four miles) over a cow pasture near Dayton. The members of the Aero Club of France, a collection of aviation enthusiasts and experimenters, were thrown into turmoil. Some accepted the claim; others were disbelieving. One of the club's leaders, attorney Ernest Archdeacon, and French oil magnate Henri Deutsch de la Meurthe had already offered a prize of fifty thousand francs to the first person who flew a one-

kilometer circuit. Now he publicly goaded the Wrights to come to France and win it. When they refused, one newspaper did not pull any punches. "The Wrights have flown or they have not flown," read the editorial. "They are in fact either fliers or liars."

Enter a wealthy Brazilian named Alberto Santos-Dumont. He had longed to fly since childhood and, as an adult living in

31

France, had been building and flying airships since the turn of the century. Turning his enthusiasm to airplanes, he studied photos of the Wright gliders and came up with the *14bis*, a collection of box kites with wheels and a propeller. It looked ungainly; in fact, it was barely controllable. But in September 1906 Santos-Dumont managed to get the *14bis* into the air, and on November 17 he flew it 722 feet. The spectators, who included members of the press invited by Santos-Dumont, were electrified at the idea that they had witnessed history's first flight; they had no reason to believe otherwise. The stories of Kitty Hawk were just that—stories—but Santos-Dumont had flown before their eyes. The diminutive Brazilian (he stood five foot five) was now a national hero. The French newspaper *Le Figaro* trumpeted, "What a triumph!… The air is truly conquered. Santos has flown. Everybody will fly."

But it would be years before anyone in Europe flew well. For the time being, French aviation was hampered by lack of knowledge. Like Santos-Dumont, many enthusiasts tried to emulate the Wrights' designs. Although some did better than the Brazilian had done, no one understood the means of controlling an airplane in roll. As a result, even the most successful of the early French airplanes, made by brothers Gabriel and Charles Voisin, flew awkwardly. For now, getting into the air, flying

in a straight line, and landing without crashing were enough of an achievement.

Such limitations didn't stop dozens of enthusiasts from joining the ranks of French aviators. The Voisins, who established the world's first work-for-hire airplane-construction company, built whatever bizarre, unflyable contraptions their customers wanted, as well as their own boxy creations. One of the first to purchase a Voisin biplane was Henri Farman, a man who epitomized the impetuous new breed of flier France was giving to the world. A former automobile racer (after a nearly fatal accident he'd abandoned that sport for flying, which he called "safer"), Farman was as ingenious as he was daring. He modified his Voisin and proceeded to win the Deutsch-Archdeacon prize in 1908. Never mind that he flew somewhat clumsily, making flat, wide turns around the pylons marking the one-kilometer course; the French press lionized him as it had Santos. Soon it was Farman who challenged the Wrights to a race in France.

Wilbur Wright had already decided to go to France, but not to race Farman. The brothers from Dayton had spent more than two years struggling to patent and market their invention. Now they had a patent for the basic aspects of their design, including the wing-warping system. The U.S. Army had, after little or no interest, finally agreed to purchase a Wright Flyer for $25,000, with a bonus of $5,000 if the

CURTISS *JUNE BUG* IN FLIGHT

Motorcycle racer Glenn H. Curtiss won the Scientific American Trophy, on July 4, 1908, by publicly flying the *June Bug* more than a mile in Hammondsport, New York. Ailerons on the *June Bug* were the key issue in the Wright brothers' patent infringement lawsuit filed against Curtiss. *The June Bug* was created by the 1905 Aerial Experiment Association (AEA), which included Curtiss, Frederick Baldwin, Alexander Graham Bell, Thomas Selfridge, and John McCurdy.

airplane's performance exceeded specifications. And a consortium of French businessmen was interested in licensing the aircraft to sell in their country. Both deals hinged on the Wrights making demonstration flights. Orville would fly at Fort Meyer in Virginia, and Wilbur would sail to France, having shipped an unassembled Flyer ahead of time. Arriving to find the Flyer badly damaged by mishandling in customs,

Wilbur settled in at a race course near Le Mans for weeks of repair work.

The French hardly knew what to make of Wilbur Wright. Unlike their own impassioned adventurers, this laconic American in his business suit and cap was the essence of calm reserve, an enigmatic fellow whose reticence only made him more mysterious. Curious bystanders stopped by daily to watch him tinkering with his machine,

WRIGHT MILITARY FLYER CRASH AT FORT MEYER

AEA member Lieutenant Thomas Selfridge died in the crash of the Wright Military Flyer at Fort Meyer, Virginia, in September 1908 when a propeller split during the flight. A seriously injured Orville Wright recovered from the accident, and the U.S. Army trials were postponed until 1909.

(ABOVE) **WILBUR WRIGHT WITH THE KING OF SPAIN**

Early in 1909 King Alfonso XIII of Spain came to Pau, France, to witness Wilbur's flights. Although he spent a moment sitting in the Flyer's passenger seat, he kept a promise to his wife that he would not fly.

(LEFT) **WRIGHT STUNS FRENCH ONLOOKERS, 1908**

Amazed French spectators celebrate Wilbur Wright's obvious mastery of the skies in the summer of 1908. While European aviators could barely master flying in a circle at this time, Wilbur executed turns and figure eights with ease.

WRIGHT MILITARY FLYER AT THE MUSEUM

The U.S. Army's first aircraft, a Wright 1909 model, hangs in the National Air and Space Museum. The Wrights returned to Fort Meyer in June 1909 with this improved machine, which the War Department purchased for $30,000.

many wondering whether he had flown as he'd claimed. On August 8 their doubts were resolved. That day the weather was ideal, and Wilbur decided to take the machine up for a spin. When he did, late in the clear afternoon, the assembled onlookers were reduced to awe. In a feat he would repeat more than a hundred times at Le Mans, Wilbur flew precisely and masterfully, making a series of tight, steeply banked turns, the likes of which the French had never seen. Now there could be no question that the Wrights had done everything they claimed. "Wright is a genius," said one witness. "He is the master of us all." But the Wrights' supremacy would soon be challenged by French fliers, including the man who spoke those words, Louis Blériot.

Even as France showered Wilbur with adulation, French aviators longed to equal and even surpass his achievements. One goal beckoned more than any other: to fly the twenty-two miles separating Calais and Dover, across the English Channel. In 1907 Lord Northcliffe, publisher of London's *Daily Mail*, had offered £1,000 for the first Channel flight. Then, it had seemed an impossible goal—but not so in the wake of Wilbur's triumph at Le Mans. Lord Northcliffe renewed his call in August 1908, hoping the winner would be an Englishman. As it turned out, the two airmen who vied for the honor the following July were both French.

Hubert Latham, born into a wealthy family of ship owners, was a self-described man of the world who had hunted lions in Africa, explored Indochina, and raced automobiles and speedboats in France. In the winter of 1909 he turned to airplanes. The dashing Latham was not universally admired; one aviation writer commented, "He possesses all the qualities and faults of a spoiled child. He likes praises, he adores eulogies . . . to bring glory to himself with the aeroplane, that is what Hubert Latham looks for."

Nevertheless, Latham was a superb pilot who had already won prizes. When he announced his intention to fly the Channel, he became the favorite. His airplane was nothing like the sturdy, well-trussed biplanes of the Wrights and their imitators. The Antoinette IV, designed by Léon Levavasseur, was a monoplane, fast and lightweight. Almost birdlike in its sleekness, it was nothing less than the vanguard of French aviation's new era. Unfortunately, its small engine was temperamental and had a tendency to conk out in damp weather—and there was a lot of that at Calais. When Latham wasn't grounded by rain or mist, he was struggling with his balky engine. Finally, on July 19 Latham made his attempt. A crowd of admirers watched the Antoinette shoot from the cliffs out over the ocean and disappear into the fog. He was not yet halfway to his goal when the Antoinette's engine sputtered and

(OPPOSITE, TOP) **HUBERT LATHAM AFTER RESCUE**
With his characteristic cigarette, Parisian Hubert Latham looks none the worse for wear after his rescue from the English Channel in July 1909. Latham sought the London *Daily Mail* prize for crossing the Channel in a heavier-than-air machine.

(OPPOSITE, BOTTOM) **LATHAM BEFORE TAKEOFF**
Latham's mount, the sleek Antoinette, was equipped with an unreliable engine. The liquid-cooled, largely aluminum V-8 power plant had a fuel-injection system that tended to clog in mid-flight. Luckily for Latham, the 1909 Antoinette floated better than it flew.

died. Calmly, Latham put the plane down amid the Channel's gentle swells. His rescuers found him sitting on his floating craft, casually smoking a cigarette.

Latham returned to Calais determined to try again, but fate handed the opportunity to Louis Blériot, a true underdog. An engineer-turned-businessman who had made a fortune selling accessories for the thriving automobile industry, Blériot had for years been one of France's most ardent aviation enthusiasts. His dark, Gallic features were dominated by a flowing mustache beneath a birdlike beak of a nose that seemed entirely fitting for a man who wanted to fly. His approach to airplane design had been more than a little haphazard, and his flying far from expert. Blériot had suffered many crash landings and near disasters. And by July 1909, after devoting himself to flying for two years, he faced bankruptcy, having spent not only his entire fortune but his wife's dowry. At the last minute he was able to borrow another twenty-five thousand francs, and he brought his newest craft to Calais. The Blériot XI was neither as powerful nor as graceful as the Antoinette, but it had one great advantage: its small, noisy Anzani engine. Although it routinely discharged oil and clouds of smoke, the Anzani was actually more reliable than the engine in Latham's Antoinette.

Just after sunrise on July 25, Blériot departed. A few miles away, a sleeping Latham and his assistants were startled awake by the noise of their rival's engine. Latham rushed to his new plane, the Antoinette VII, but by now the wind was stronger than the craft was designed to handle. Latham's chance was gone; his usual composure gave way to tears.

Meanwhile, over the Channel Blériot sped onward, surprised at his own calm in what he later called this "supreme" moment. Ten minutes after takeoff he was out of sight of France, alone above the Channel's wide expanse of wind-stirred waves. Before the flight he'd had to ask which direction to fly to reach Dover; now he wondered if he might be off course. After ten long minutes he spotted, he later wrote, "a gray line emerging from the sea. . . . It was the English coast." But Dover was nowhere in sight. "The wind and fog caught me. I fought with my hands, with my eyes. . . . Where the devil was I?"

Unbeknownst to Blériot, the wind had blown him to the north. Now he spotted three ships he presumed were heading toward port; turning south to follow them, he flew past the famous white cliffs. He could hear cheers from sailors on the decks; he wished he could have asked them for directions, but he spoke no English. Moments later the Dover castle came into view. "A mad joy comes over me. . . . I am over land!" As he neared the ground, Blériot could see French journalist Charles Fontaine "desperately waving a tricolor,

LOUIS BLÉRIOT

Auto headlight manufacturer Louis Blériot was Latham's chief competitor. Twice burned by his monoplane's exhaust system and strapped for cash, Blériot was determined to win as many prizes as possible, including the *Daily Mail's.*

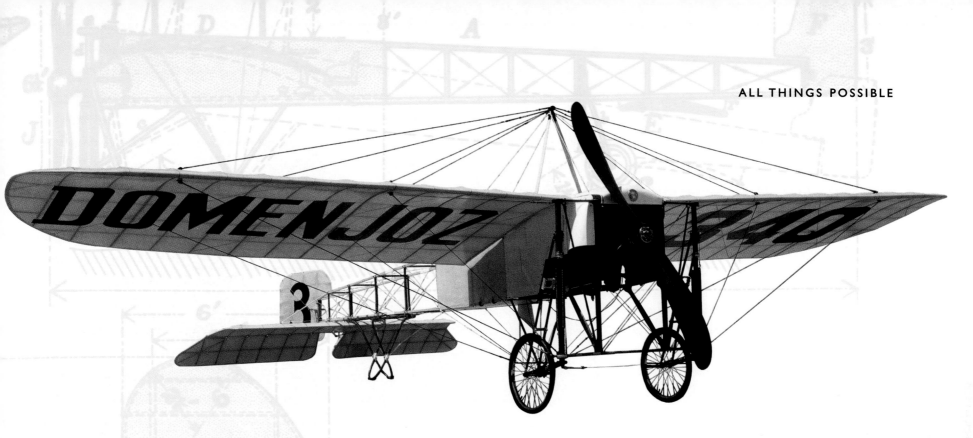

(ABOVE) **BLÉRIOT XI**

The National Air and Space Museum's 1914 Blériot XI is a monoplane like its illustrious parent, the cross-Channel Blériot. The reliable air-cooled rotary engine produced great clouds of blue smoke and bathed the pilot in castor oil lubricant.

(LEFT) **BLÉRIOT CHANNEL CROSSING**

A contemporary postcard celebrates Louis Blériot's July 25, 1909, flight across the English Channel. "It is a strange position, to be alone, unguided, without compass, in the air over the middle of the Channel," Blériot later wrote. "I touch nothing, my hands and feet rest lightly on the levers. I let the aeroplane take its own course."

41

out alone in the middle of a field, bawling, 'Bravo! Bravo!'" Blériot headed for him and then, fighting the wind, made a landing so rough that it damaged the craft's landing gear and propeller. But that hardly mattered; he was unharmed, and he had done it.

In London, and then on his return to Paris, the public responded with a hero's welcome. The reaction hinted that the real significance of Blériot's flight was not measured in distance—Wilbur Wright had flown many times farther at Le Mans—but in cultural impact. The airplane had become a bridge between nations; suddenly, the possibility of air travel was real. So was the prospect of aerial warfare. In London newspaper editorials foretold the end of Britain's proud isolation, of an age when mighty sea power would be rendered impotent by high-flying airplanes. But in France thoughts were far more joyous. Blériot's flight stirred French pride, not only as a national victory but as a harbinger of future achievements. As French inventor Clément Ader had said, "Whoever will be master of the sky will be master of the world."

The Channel crossing was the first great achievement of 1909; the second came a month later, when thirty-eight fliers converged at Reims for the world's first air meet. If anyone needed evidence that the aviation age had arrived and that France

was its capital, he or she had only to witness the swarm of airplanes buzzing over the thousands of spectators. Antoinettes, Blériots, Farmans, and Wright models took to the air, alone and together, in an exhibition of aerial progress. All sorts of new designs—monoplanes and biplanes, propellers in front (tractors) or in back (pushers), water-cooled engines and air-cooled ones—competed in races of distance and speed. At times there were as many as seven machines aloft at once, filling the sky with wood and fabric. And before long, the ground was littered with the wreckage of flimsy, temperamental craft whose pilots had turned too sharply, climbed too steeply, or crossed paths with the devils of air turbulence. Fortunately, the crashes claimed no casualties. Even Blériot him-

AIRCRAFT ROUNDS PYLON IN RACE

An Henri Farman military biplane rounds a pylon during a 1912 air race. For thousands of people, their first glimpse of an airplane came at these contests. European machines soon equaled or bettered their American counterparts in performance.

self, whose Type XII monoplane burst into flames because of a leaky fuel line, narrowly escaped serious injury.

None of that lessened the importance of Reims as a watershed moment for European aviation. After the event's critical and financial success, air meets boomed in popularity. For promoters, there was the potential for enormous profits, and for a thrill-seeking public, a new kind of excitement as men and airplanes strained against their limits. Above all, the meets were a proving ground for aviation ad-

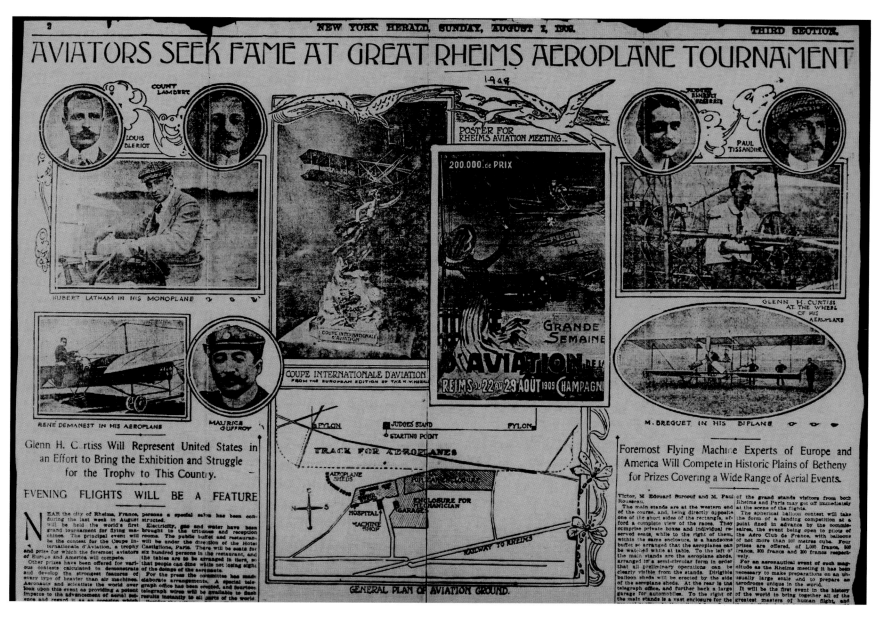

AVIATORS SEEK FAME AT GREAT RHEIMS AEROPLANE TOURNAMENT

HUBERT LATHAM IN HIS MONOPLANE

RENÉ DEMANEST IN HIS AEROPLANE

MAURICE GUFFROY

POSTER FOR RHEIMS AVIATION MEETING

COUPE INTERNATIONALE D'AVIATION FROM THE EUROPEAN EDITION OF THE N. Y. HERALD

200.000 de PRIX GRANDE SEMAINE D'AVIATION REIMS du 22 au 29 AOÛT 1909 CHAMPAGNE

PYLON JUDGES STAND PYLON STARTING POINT

TRACK FOR AEROPLANES

AEROPLANE SHEDS POPULAR ENCLOSURE ENCLOSURE FOR MECHANICIAN HOSPITAL GARAGE MACHINE SHOP RAILWAY TO RHEIMS

GENERAL PLAN OF AVIATION GROUND.

GLENN H. CURTISS AT THE WHEEL OF HIS AEROPLANE

M. BREGUET IN HIS BIPLANE

PAUL TISSANDIER

Glenn H. Curtiss Will Represent United States in an Effort to Bring the Exhibition and Struggle for the Trophy to This Country.

EVENING FLIGHTS WILL BE A FEATURE

NEAR the city of Rheims, France, persons during the last week in August will be held the world's first grand tournament for flying machines. The principal event will be the contest for the Coupe Internationale d'Aviation, a trophy and prize for which the foremost aviators of Europe and America will compete.

Other prizes have been offered for various contests calculated to demonstrate and develop the strongest features of every type of heavier than air machines. Aeronauts and scientists the world over look upon this event as providing a potent impetus to the advancement of aerial science and regard it as an occasion which...

Foremost Flying Machine Experts of Europe and America Will Compete in Historic Plains of Betheny for Prizes Covering a Wide Range of Aerial Events.

NEWSPAPER COVERAGE OF REIMS AIR RACES

The August 1909 Reims races lasted a week and included thirty-five entries—among them Henri Farman, Louis Blériot, Hubert Latham, and Glenn Curtiss. World attention focused on the records set, poor weather, and numerous nonfatal crashes.

vances. Special events were designed to test the capabilities of pilots and machines: there were competitions for speed, flight duration, quickest takeoff, sharpest turns, and greatest altitude. For the fliers themselves, the risks of competition were compensated by the lure of fame and fortune. But their aircraft were not always up to the

challenge. In August 1910 a Paris-born Peruvian named Jorge Chávez climbed to almost 8,500 feet in his Blériot XI, far higher than anyone before him and nearly twenty times the record set at Reims a year earlier. The next month, with little preparation, Chávez attempted to steer his monoplane over the Alps at the Italian-

REIMS AIR MEET—CRASHED AIRPLANE

An all-too-typical landing during the 1909 Reims meet. Flying over the course, Glenn Curtiss counted a dozen disabled machines below him. Nearly 90 flights out of 120 exceeded three miles, however, proving that aviation had come a long way in six years.

(RIGHT) **GRANDE QUINZAINE D'AVIATION**

Poster for the Grande Quinzaine d'Aviation features a speeding Antoinette. As prize purses grew, long-distance races spurred aviators to increase their piloting skills, and manufacturers to build better machines.

(BELOW) **CIRCUIT AÉRIEN INTERNATIONAL**

Not to be outdone by the French, Italy's Società Italiana di Avizione invited fliers to compete on the plains of Lombardy nine days after Reims closed. Glenn Curtiss won the speed contest at both meets.

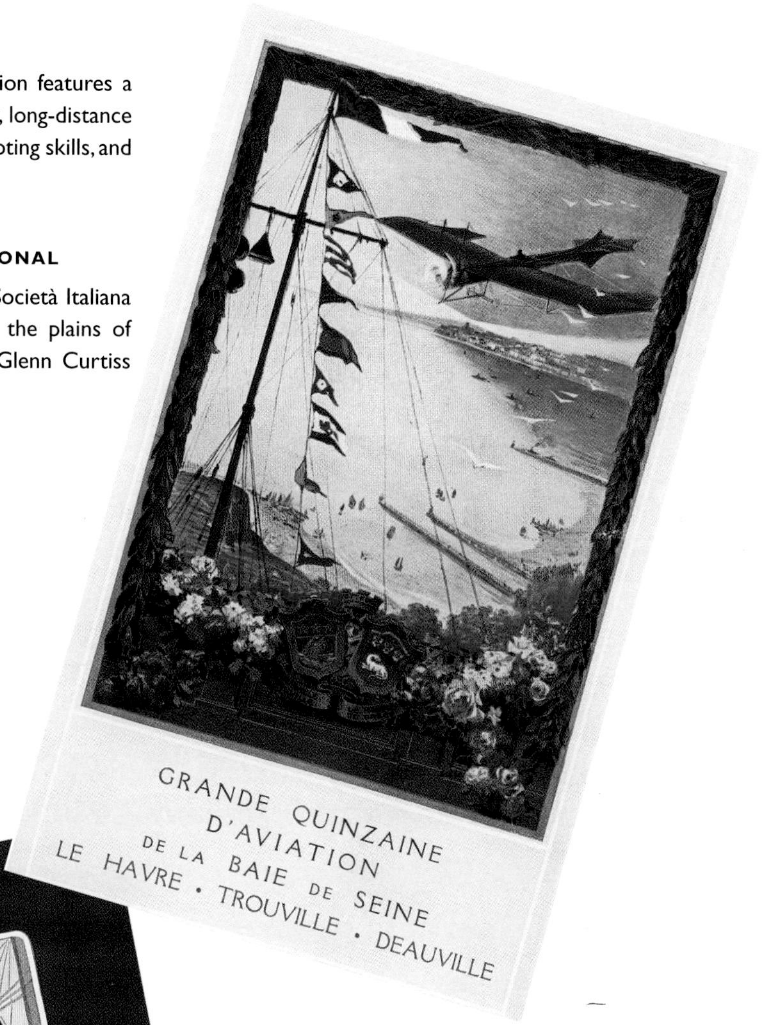
GRANDE QUINZAINE
D'AVIATION
DE LA BAIE DE SEINE
LE HAVRE • TROUVILLE • DEAUVILLE

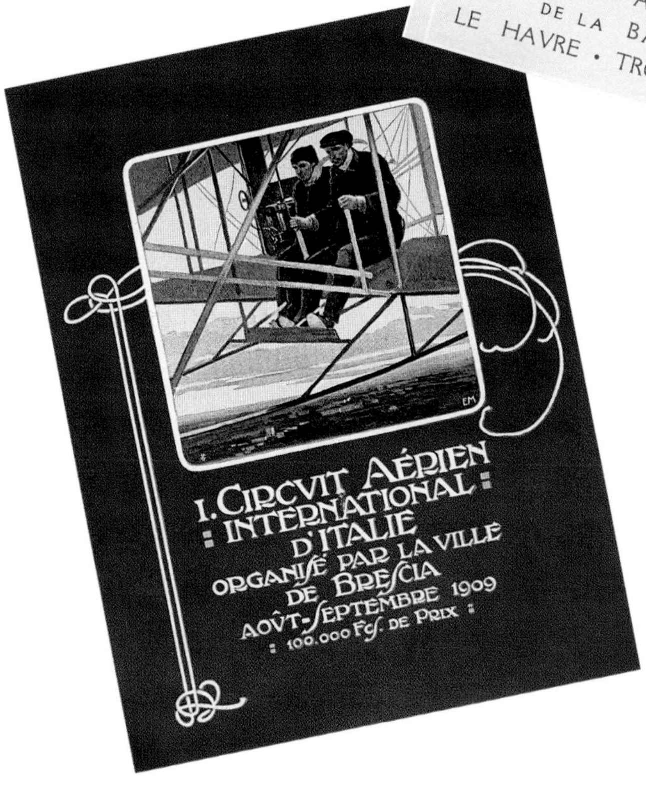
I. CIRCVIT AÉRIEN
INTERNATIONAL
D'ITALIE
ORGANISÉ PAR LA VILLE
DE BRESCIA
AOÛT-SEPTEMBRE 1909
100.000 Fcs. DE PRIX

Swiss border. Tossed by violent winds, he struggled past towering peaks to emerge high over Italy, apparently headed for a safe landing. But unknown to Chávez, the flight had dangerously weakened the wings; at almost the last instant, they snapped back against the fuselage, and the Blériot plummeted to Earth. Chávez died days later from his injuries. His celebrated final words: "Higher, always higher." A week later, even as critics decried Chávez' loss, crowds flocked to an air meet in Milan, Italy. Not even death would stem the enormous popularity of these aerial contests.

Meanwhile, lavish cash prizes spurred aviators to bridge cities by air. In April 1910, when Frenchman Louis Paulhan flew from London to Manchester—besting England's popular Claude Grahame-White to win £10,000 from the *Daily Mail*—he did so by following a specially hired train. By the following year, aviators' navigational skills had improved enough for promoters to stage a slew of great races that spanned entire nations, including circuits of Britain, Germany, Belgium, and even Europe itself. Endurance was the key, for the pilots every bit as much as their machines, and no one demonstrated this more than American Cal Rodgers. In September 1911 the former college football star took off from New York's Sheepshead Bay determined to cross the United States by air within thirty days. If successful, he would capture not only a record for distance but the $50,000 in prize money offered by publishing giant William Randolph Hearst. Forty-nine days and 4,231 miles later—after enduring equipment problems, crash landings, bad weather, and even souvenir hunters who stole pieces of his biplane, the *Vin Fiz*—Rodgers reached Pasadena, California. The final twenty miles to Long Beach were the hardest: along the way Rodgers crashed on takeoff and ended up in the hospital with both legs and his collarbone broken and a concussion. "I am going to finish that flight," declared Rodgers from his hospital bed. And he did on December 10, eighty-four days after the journey began. The following April, returning to Long Beach for an exhibition flight, an exuberant Rodgers dove into a flock of seagulls, one of which became wedged in the tail mechanism, causing Rodgers to crash fatally. The epitaph on his tombstone reads, "I endure—I conquer."

But in the quest to conquer the air, the year 1911 belonged to France, whose fliers held all major records for speed, distance, endurance, and altitude. By contrast, of all the nations that had learned to fly, the United States lagged far behind. Americans had issued by far the fewest number of pilot's licenses. No American had won an air race since Glenn Curtiss's 1909 victory at Brescia, Italy—nor would any for years to come. With relatively few air

(RIGHT) *VIN FIZ* AT THE MUSEUM

The Wright EX *Vin Fiz*—named for the soft drink advertising it carried—hangs in the National Air and Space Museum. In April 1912, Rodgers died in a crash near the spot where he had finished his record flight only four months earlier.

(OPPOSITE, TOP) CAL RODGERS WITH THE *VIN FIZ*

With his characteristic cigar and smile, American daredevil Calbraith Perry Rodgers prepares for takeoff. Flying a 1911 Wright EX, Rodgers sought William Randolph Hearst's prize of $50,000 for the first flight across the United States in thirty days. The Wright brothers did not think such a feat was possible.

(BELOW, AND OPPOSITE, BOTTOM) *VIN FIZ* CRASH

AND REPAIRS

This is how Rodgers's aircraft looked during more than half of its forty-nine-day trip across the country. He crashed nineteen times during his trip. Even with help from Wright master mechanic Charles Taylor and financial support from the Armour Company, the Hearst prize remained out of reach.

(ABOVE) CAL RODGERS'S JOURNEY

This map shows the route of the *Vin Fiz* during its trouble-plagued transcontinental flight. Rodgers made sixty-nine stops along the way, finishing eighty-four days after he began—with a total of 82 hours and 4 minutes in the air.

47

meets in the United States, most activity centered on roving troupes of stunt fliers. One was Lincoln Beachey, who became aviation's greatest daredevil with an assortment of loops and rolls and high-altitude, unpowered dives that made spectators hold their breath. But not even Beachey's hair-raising stunts could dispel the nationwide apathy toward aviation. America, just now emerging as a world power, did not face the threat of war that had hung over the European nations since the turn of the century. That, as much as anything else, explained the Army's lack of interest in the invention of Wilbur and Orville Wright. The inventors themselves, embroiled in patent suits against rival Glenn Curtiss, had ceased putting their energies into improving their creation. The nation that had invented the airplane was letting other countries lead the way.

In Europe aviation got a boost not only from flying clubs and wealthy enthusiasts but from governments. In England, Germany, and Russia, new research centers were inaugurated to probe the mysteries of how and why airplanes flew. New engineering facilities, including England's Farnborough, were created to explore advanced methods of airplane construction. And in England and Germany, whose governments sensed the approach of war, special attention was given to the airplane's military potential.

The fruits of this research were evident

(LEFT AND ABOVE) **U.S. AIR MEET EPHEMERA**
Ephemera from the National Air and Space Museum's collections recall the first glory days of aviation in the United States—a program from the 1911 Harvard-Boston Air Meet and a ticket from the 1910 Belmont Park "Aviation Tournament."

(BELOW) **LINCOLN BEACHEY RACING BARNEY OLDFIELD**
Stunt pilot Lincoln Beachey battles racer Barney Oldfield and proves flying is faster than driving. Formerly a dirigible pilot, Beachey was renowned for his loops, spirals, and dives. He died in 1915 when the wings snapped off his new monoplane during a dive.

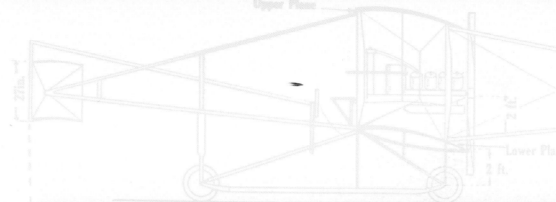

(LEFT) CURTISS MODEL D "HEADLESS PUSHER"

A favorite of stunt fliers like Lincoln Beachey, the Curtiss D-III headless biplane (also depicted here in line drawings) had several remarkable features. It was easy to assemble and disassemble for shipment or repairs and had bamboo tail struts, which were light and splinter-resistant. The National Air and Space Museum's Curtiss, shown here after restoration, was built by Glenn Curtiss after World War I.

(BELOW, LEFT) HARRIET QUIMBY, 1912

The first woman licensed to fly in the United States, Harriet Quimby earned another first by becoming the first woman to fly across the English Channel in April of 1912.

(BELOW) LOS ANGELES AIRSHOW, 1910

This program from the National Air and Space Museum's collection is from the first U.S. International Air Meet at Los Angeles, California, in January 1910. Frenchman Louis Paulhan flew to an altitude of 4,164 feet for a new world record at this competition.

49

(LEFT) **JULES VÉDRINES**

Sole finisher of the 1911 Paris–Madrid air race, Jules Védrines glowers for the camera. Nearly attacked by eagles over the Pyrenees, the volatile aviator was famous for his aggressive flying and amazing endurance; he flew from Paris to Cairo in 1913.

(BELOW) **1913 DEPERDUSSIN**

French aviation led the world by 1913. The slim Deperdussin, created by Louis Béchereau, featured a 160-horsepower rotary engine; short, low-aspect-ratio wings; and a streamlined monocoque fuselage. Engine torque made the aircraft hard to fly, but at speeds in excess of 100 miles per hour all but one pilot at the Coupe Internationale d'Aviation that year flew these remarkable airplanes.

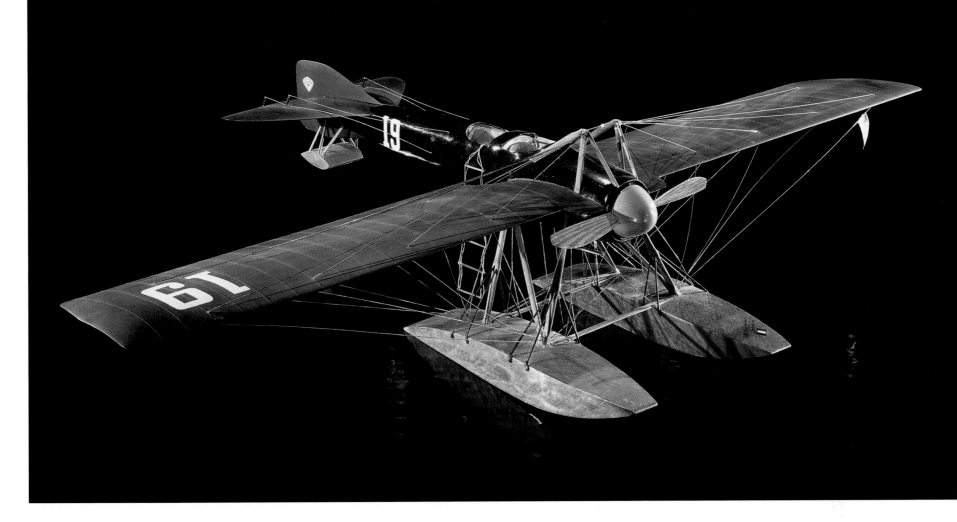

in the quickening pace of improvements, as new craft broke records for distance and altitude. The problem of endurance was being surmounted; the great races had made that clear. If airplanes were to be effective in warfare—and if they were to win such celebrated prizes as the Gordon Bennett trophy—they would have to perform at ever higher speeds.

Designers in France made some of the most significant advancements. In 1911 a young French engineering graduate named Edouard de Niéport took two key

steps to reduce drag: enclosing his monoplane's entire fuselage in fabric and putting a cowling around the engine. But the most impressive strides were made by engineer Louis Béchereau and his colleagues. Their 1912 Deperdussin racer represented a giant leap toward the modern airplane. Instead of a wooden, fabric-covered frame, the Deperdussin's fuselage was a strong shell made of stressed plywood. This *monocoque* construction (from the French word for "shell") also saved weight by eliminating the need for internal

MODEL OF 1913 DEPERDUSSIN FLOATPLANE

This scale model at the National Air and Space Museum captures the lines of the 1913 Deperdussin floatplane. A craft like this, piloted by Maurice Prévost, won the Inaugural Schneider Cup race for seaplanes. Competitors in this contest were required to navigate two and a half nautical miles on their floats to qualify.

51

THE MUSEUM'S FOWLER-GAGE BIPLANE

This biplane, built by the Gage-McClay company, was flown by early aviation pioneer Robert Fowler. On April 27, 1913, Fowler piloted the airplane, with a passenger, on the first aerial crossing of the Isthmus of Panama. The aircraft was later used to inspect telegraph lines in California.

(LEFT) **ADOLPHE PÉGOUD FLYING A LOOP-THE-LOOP, 1913**

Numerous accidents with early monoplanes led many to doubt their safety. Blériot's chief demonstration pilot, Célestin Adolphe Pégoud, put most of those doubts to rest in September 1913. Pégoud's loops, vertical S, tail slide, and inverted flying impressed even his fellow pilots. In the wake of this success, he formed the first aerial stunt team in October.

(BELOW) **FIRST PARACHUTE JUMP FROM AN AIRPLANE**

On March 1, 1912, American Captain Albert Berry made the first parachute jump from an airplane, 1,500 feet above Jefferson Barracks in St. Louis. The parachute was stored in a container attached to the aircraft.

bracing. With a 160-horsepower Gnôme engine, the Deperdussin was the fastest thing in the sky. In mid-January 1912 expert pilot Jules Védrines achieved the unprecedented speed of 100 miles per hour. Before the year was out, he would pilot the Deperdussin to six world speed records. The airplane was now by far the swiftest vehicle in existence.

With each passing day, it seemed, new exploits demonstrated aviation's potential to change the world. The emergence of stunt flying—typified by Adolphe Pégoud, who performed the first loop-the-loop in public—showed that the airplane was becoming capable of sophisticated maneuvers that foreshadowed aerial combat. And the airplane now took to the water, thanks largely to the efforts of American flier and inventor Glenn Curtiss. Carrier aviation had gotten its start late in 1910, when Eugene Ely took off from the deck of the cruiser USS *Birmingham*. Two months later he steered his Curtiss biplane to a shipboard landing on the cruiser USS *Pennsylvania*, stayed long enough to have lunch with the ship's captain, and took off again. Since then Curtiss had developed the first practical seaplanes, including an amphibian model called the Triad that could be converted to land operations by lowering a special set of wheels.

By April 1913 seaplanes merited their own competition, the Schneider Trophy Race, held in Monaco. Curtiss seaplanes were flying for the U.S. Navy, and the navies of Germany, Russia, and Japan. At the same time the flying boat—another Curtiss innovation, in which the fuselage acts as a hull to provide flotation—became popular with sportsmen. And in Florida as 1914 opened, the Saint Petersburg–Tampa Air Boat Line offered an alternative to road-weary passengers with a twenty-three-minute hop across Tampa Bay in a single-passenger flying boat made by Curtiss's competitor Thomas Benoist. As America's first scheduled airline service, it made 1,200 trips during the 1914 tourist season.

But it was in Saint Petersburg, Russia, that one of the decade's greatest flights began in mid-August 1914. There aviation pioneer Igor Sikorsky had been creating the largest airplanes in existence. Convinced that single-engine aircraft were too unreliable, he had built a four-engine biplane called the *Grand* in 1913. Despite being underpowered—it cruised at only fifty miles per hour—the *Grand* was able to remain aloft for more than six hours,

FIRST CARRIER LANDING, 1911

In 1910, a demonstration by Eugene Ely in Hampton Roads, Virginia, proved the feasibility of carrier aviation and sparked the U.S. Navy's interest in flight. Taking off from a wooden platform built aboard the cruiser USS *Birmingham*, Ely landed safely two miles away. In 1911 he did just the opposite (shown above), landing on the USS *Pennsylvania* in San Francisco Bay.

SCHNEIDER TROPHY RACE IN MONACO, 1913

English designer Thomas Sopwith's seaplane is shown here at Monaco in 1913. His Pup, Camel, and Snipe fighter designs contributed greatly to the Allied effort in World War I, and he would later be knighted for his work.

(ABOVE) ECKER FLYING BOAT

The National Air and Space Museum's 1912–13 Ecker Flying Boat came from a TV repair shop attic in Syracuse, New York, in 1961. Herman Ecker built his airplane from a boat hull and motor, using carpet tacks to secure the wing fabric and motorcycle spokes as turnbuckles for the bracing wires.

(RIGHT) SAINT PETERSBURG–TAMPA AIRBOAT LINE, 1914

On the tenth anniversary of the Wright brothers' first flight, the city of Saint Petersburg, Florida, contracted with Thomas Benoist to operate a regular airboat line to Tampa. The Saint Petersburg–Tampa Airboat Line, organized by Paul Fansler (second from right), operated for three months and carried 1,204 passengers.

(ABOVE) **ZEPPELIN LZ4 AIRSHIP**

Count Ferdinand von Zeppelin's LZ4 airship rests on the waters of Germany's Lake Constance in this postcard depiction. The German public was so enamored of the giant airships that when the LZ4 crashed in 1908, donations poured in from around the country, allowing Zeppelin's company to recover from the misfortune.

(ABOVE, LEFT) **"COME TAKE A TRIP IN MY AIR-SHIP"**

This colorful sheet music cover from the Museum's collection depicts an early dirigible design of the type flown by A. Roy Knabenshue and Captain Thomas Baldwin, with a nonrigid envelope and guide ropes for maneuvering it on the ground.

(LEFT) **LEBAUDY AIRSHIP WITH EIFFEL TOWER IN BACKGROUND**

M. Lebaudy pilots his dirigible *Le Jaune* toward the Eiffel Tower on November 20, 1903. Less spectacular than their heavier-than-air cousins, powered airships—pioneered by the likes of Santos-Dumont and Zeppelin—would play an important role in the upcoming World War.

breaking world endurance records. But Sikorsky's crowning achievement was the *Il'ya Muromets*, a five-ton behemoth with a wingspan of just over 100 feet. The spacious cabin, appointed with enough chairs and sofas to seat sixteen, as well as electric lights, a bedroom, and a toilet, was downright luxurious compared with the spartan conditions on most aircraft of the day. And on August 20, 1914, Sikorsky flew the *Il'ya Muromets* from Saint Petersburg to Kiev and back, a journey of nearly 1,600 miles that anticipated the age of luxury air travel.

But even before this impressive flight could begin, it was clear that the *Il'ya Muromets* would soon have a use its creator hadn't intended. On August 19 the heir to the throne of Austria-Hungary was assassinated in the Serbian city of Sarajevo, touching off a world war that would be remembered as being among the bloodiest years in history. During World War I a burgeoning aviation industry would incorporate the phenomenal progress made during the six years after Wilbur Wright flew at Le Mans, setting the stage for even greater advances. In the meantime, aviation—no longer the sole province of inventors and enthusiasts, sportsmen and daredevils—would show a newer, darker face. ■

(ABOVE) *IL'YA MUROMETS*

Russian aviation pioneer Igor Sikorsky made astonishing strides building giant multi-engine aircraft at a railroad car factory outside Saint Petersburg in 1913. The *Il'ya Muromets* (pictured above) had four engines, two balconies, and compartments with electric lights and heat.

(OPPOSITE) *IL'YA MUROMETS* IN FLIGHT

Artist James Dietz depicted *Il'ya Muromets*' 1914 flight over Saint Petersburg with sixteen passengers aboard. It later was flown for over six hours, and Sikorsky himself took it on a 1,600-mile round trip. This amazing aircraft would later serve the czar as the world's first long-range bomber.

CHAPTER THREE

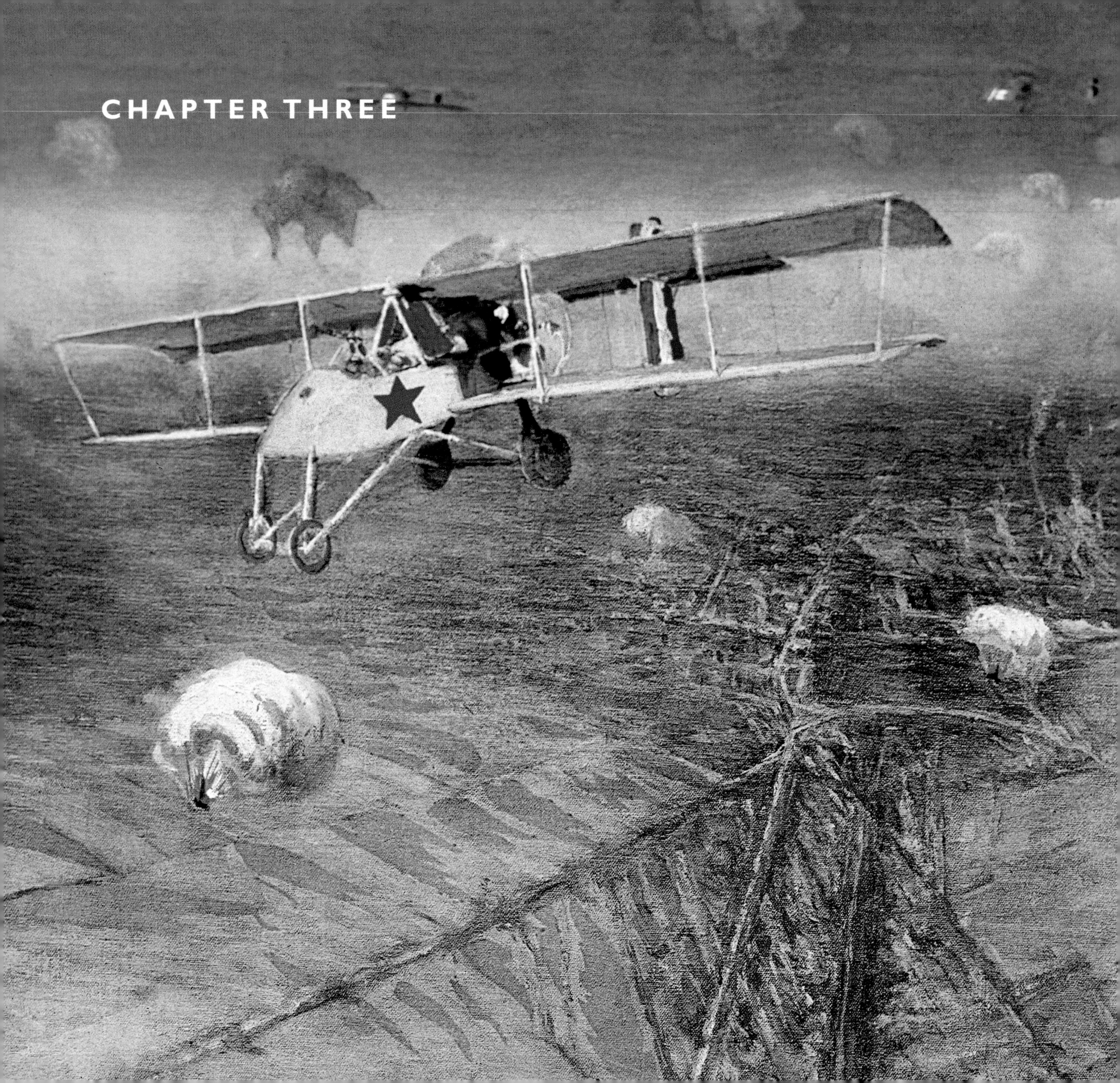

A WAR IN THE AIR

A new and terrible form of warfare exploded across Europe in August 1914. It wasn't just the scale of World War I that was unprecedented, as coalitions of nations mobilized vast citizen armies, it was the ferocity. The technology of mechanized warfare, epitomized by the machine gun, turned the battlefields of Europe into killing fields. In a matter of weeks World War I took on a character no one had anticipated. Instead of the brief clash—weeks or months at most—predicted by many, the "war to end all wars" became the most stalemated conflict in history. The deadlock came about, in part, because of an invention that only visionaries had considered a weapon: the airplane.

As the armies of Germany's Kaiser Wilhelm marched toward Paris late in August 1914, lone scouts of the city's aviation units flew high overhead, gathering information on enemy troop movements to bring back to battlefield commanders. Before the war, despite the extravagant claims of aviation proponents, some generals on both sides had doubted the military value of the airplanes that their governments had purchased. One British commander declared flatly that for reconnaissance, the airplane would never replace cavalry. It may be understandable,

then, that when a French pilot reported to Paris that the advancing Germans had turned, exposing their flank to attack, the intelligence chief didn't believe it.

Fortunately, General Gallieni, the commander of the Paris fortifications, believed in the airplane's potential and trusted the pilot's report. Consequently, French forces managed to rout the German attack and save their city in the First Battle of the Marne. In the battle's wake, mobile warfare rapidly gave way to trench warfare, as both sides dug in for a war of attrition. Similar events shaped the fighting on the war's eastern front. At the Battle of Tannenberg, outnumbered German armies managed to repel the attacking Russian forces with help from German aviators who provided a valuable overview of the fighting. The advent of aerial reconnaissance made surprise attacks impossible. No other use of the airplane had more impact on the war than the ability to observe from the air.

No wonder, then, that both sides considered the enemy's observation planes prime targets. With few antiaircraft weapons available at first, the pilots themselves had to take action; when airborne scouts from opposite sides encountered each other, aerial combat was born. At first they shot at each other with pistols, mostly unsuccessfully, or rammed their airplanes—a practice that was more effective but sometimes fatal for both fliers. Then, in the spring of 1915, came the change that began to transform the airplane itself into a deadly weapon. Roland Garros, a legendary French flier who in 1913 had been first to cross the Mediterranean by air, mounted a forward-firing machine gun on his Morane-Saulnier monoplane and fitted the propeller with protective metal guards. During the first three weeks of April, Garros shot down three German planes before he himself was downed by enemy ground fire. He had become the world's first true fighter pilot, and when his name appeared in military dispatches, the French press transformed Garros the airman into Garros the war hero.

Such adulation broke with established military policy against singling out individuals. But the public welcomed news of

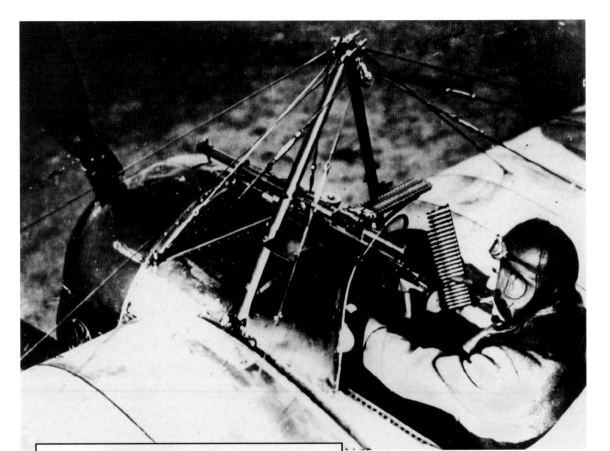

ROLAND GARROS

Frenchman Roland Garros (shown at left) was the first to fly across the Mediterranean, in 1913. He revolutionized aerial combat in 1915 by mounting a machine gun on his Morane-Saulnier Type L monoplane. Armored deflectors similar to those shown on this Type N above kept him from shooting off his own propeller, but the Germans learned his secret when they captured Garros and one of his fighters in April.

JEANNIN *STAHLTAUBE* MODEL

This scale model captures the delicate lines of the Jeannin *Stahltaube* (steel dove), a German monoplane scout from 1914 to 1916. The aircraft's Mercedes water-cooled, six-cylinder engine produced 120 horsepower and gave it a speed of sixty-two miles per hour. This two-seater could reach an altitude of 8,700 feet and was used for aerial reconnaissance work. Aviators flying the *Stahltaube* spotted the Russian advance into East Prussia, allowing the German army to envelop the Russians at the Battle of Tannenberg for a decisive defeat.

WORLD WAR I ACES

Romanticized in postwar years, the first fighter pilots were often lone hunters with eccentric personalities and deadly marksmanship skills. Early fighter aircraft were scarce and group tactics nonexistent. The Germans were the first to have an effective fighter—Fokker's *Eindecker*—and the French responded by grouping their fighter pilots together. On all fronts, these "knights of the air" led short, violent lives.

RENÉ FONCK (FRANCE)

MICK MANNOCK (ENGLAND)

the military aviators, and as the conflict dragged on, governments realized that the fliers' exploits were diverting attention away from the horrors of the ground war. In the midst of unprecedented carnage— an estimated 11 million dead and nearly twice as many wounded in four years— these heroic young men seemed literally and figuratively above it all. In the mind of the public, the pilots were a chivalrous lot who enjoyed a life of adventure safely removed from the trenches, the barbed wire,

machine guns, and poison gas. Even the foot soldiers saw the aviators as a lucky breed. "I envied the fliers," wrote one American infantryman who went on to become a pilot. "The other fellows were sailing around in the clean air while I had to duck shells all the time and run chances of being caught by the machine guns and snipers. Of course, the aviators were also being shelled, but they never seemed to get hurt."

That was a huge misperception. During

the war, pilots faced the same chances of being injured or killed as did their comrades on the ground, often for reasons that had nothing to do with the enemy. Poor training did in many student pilots before they ever saw combat; malfunctions in flight killed many more. Even when airplanes worked, they could be devilishly hard to fly. If an aviator wasn't battling torque from his rotary engine, he was struggling to calculate his position with map and compass. On top of everything, of

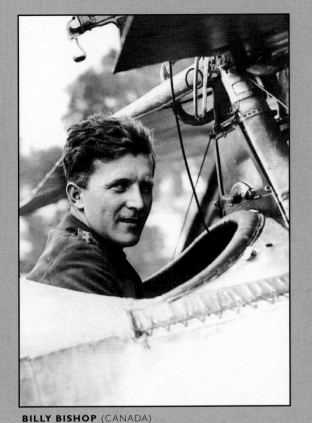

BILLY BISHOP (CANADA)

FRANCESCO BARACCA (ITALY)

MAX IMMELMANN (GERMANY)

course, he had to keep an eye out for the enemy and try to shoot him down. "The man at the controls of a fighter," said one French flier, "is pilot, machine gunner, and observer all rolled into one. It's no picnic."

When a fighter pilot had mastered this demanding art, his troubles were just beginning. The deafening engine noise, the sickening odor of hot castor oil from the engine, and the constant blast of a gale-force slipstream added up to an assault on the senses. Fliers patrolling the skies above

ten thousand feet faced frigid cold and lack of oxygen. (As a result, they dressed like polar explorers; the silk scarf that would become a symbol of aerial bravery was, above all, for keeping warm.) Blocked ear canals often made descents painful. And during violent maneuvers—for example, in the heat of a dogfight—pilots were subjected to airsickness and crushing acceleration.

Such physical hardships paled before the mental and emotional demands of aerial combat. The view from the roof of the

world was spectacular, but pilots could ill afford to let their attention wander. The enemy was almost always lurking—above, below, or somewhere in the distance. (Kills were often scored by stealth; many pilots never saw their attackers.) Survival might mean being able to spot an opponent a mile away. But no one could hope to score from a distance, and inevitably pilots were drawn into confrontation. Those brief minutes of what one aviator called "concentrated violence" demanded a kind of

65

OSWALD BOELCKE (GERMANY)

GODWIN BRUMOWSKI (AUSTRIA-HUNGARY)

ALEXANDER A. KOZAKOV (RUSSIA)

all-out flying beyond all prior experience. And while some later spoke of an adrenaline-charged excitement like that of prewar stunt fliers who survived a brush with death, most confessed to an emotion old to warfare but new to flying—terror.

Nothing was more terrifying than missions to strafe ground targets or attack the enemy's tethered observation balloons. If a pilot managed to avoid running into the ubiquitous telegraph wires, he descended into a storm of ground fire. Antiaircraft

shells were bad enough; bullets were worse. With their gas tanks behind their seats, their machine guns loaded with incendiary bullets, and their planes covered with highly flammable doped fabric, fliers knew a direct hit could send them down in flames. Parachutes small enough to fit in the cockpit weren't available until late in the war, and even then only German aviators had them. Many fliers carried revolvers for the purpose of committing suicide rather than become "flamers." No

wonder that the mere thought of being sent on a strafing mission was enough to send shivers through a pilot. In some English squadrons, if a man lived through his first such assignment, he was never sent on a second.

But the rewards for success were great. Those who triumphed, not once but repeatedly, won fame and adulation enough to last a century. Five or more victories qualified a pilot for the title of "ace," a term used first in France that became

(RIGHT) **MANFRED VON RICHTHOFEN** (GERMANY)

Manfred Freiherr von Richthofen—the infamous Red Baron—began the war in the infantry on the Russian front. He started his flying career as a backseater in May 1915. Assigned to Boelcke's *Jasta 2* in September, Richthofen immediately demonstrated his remarkable shooting skills, earning the Pour le Mérite in January. Von Richthofen was known to collect the serial numbers and names of his victims; he himself was killed in April 1918 with eighty confirmed kills.

(RIGHT, BELOW) **POUR LE MÉRITE MEDAL**

First awarded to Max Immelmann and Oswald Boelcke, the Pour le Mérite often came at a price. Both men received the award in 1916, and both were dead within a year.

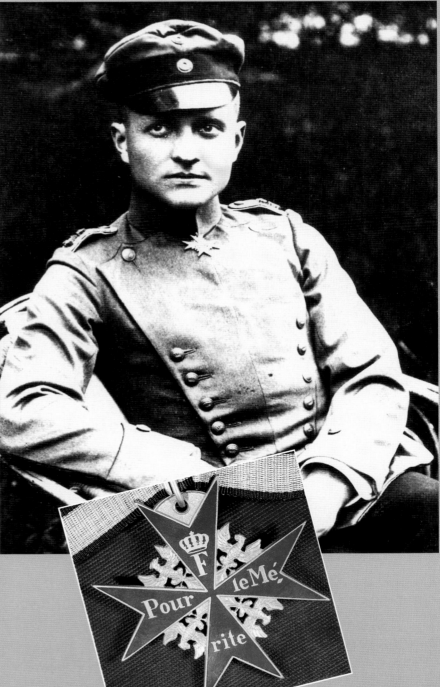

(ABOVE) **EDDIE RICKENBACKER** (USA)

Race car driver Edward Rickenbacker finished the war as America's top ace, with twenty-six confirmed victories—including five observation balloons. Rickenbacker went to France as General Pershing's staff driver before starting his flight training at the age of twenty-seven—old for a fighter pilot. He joined the 94th "Hat in the Ring" Squadron in March 1918.

Trained as a mechanic, Georges Guynemer volunteered for pilot training in 1915. Flying SPAD fighters, he scored fifty-three victories in two years, making him one of France's leading aces. Guynemer's mysterious 1917 disappearance in combat made him a martyr in France, as seen in this poster depicting him being borne to heaven by angels.

synonymous with heroism in the air. During the war, the aces' exploits became the stuff of myths. When French "ace of aces" Georges Guynemer was shot down with fifty-three victories to his credit, his name was inscribed within the Pantheon. Today the names of the aces—Germany's Max Immelmann and Oswald Boelcke, England's Mick Mannock, America's Eddie Rickenbacker, and dozens more—still endure, decades after other details of World War I have faded from memory.

For all the attention they received, the aces—and fighter pilots in general—had little effect on the war; a single bomber could destroy more enemy planes on one night's raid than an entire week of fighter missions. But fighter planes themselves underwent a remarkable transformation. At the start of the war, a fighter was, in the words of British flier Charles Illingworth, "a fragile contraption of lath and piano wire." By 1916 the pressure of the war was forcing advances in airplane design, especially in Germany. After the United States entered the war in 1917, the Germans, desperate to overcome the Allies' numeri-

(RIGHT) **DAREDEVIL ACES**

Comic books and movies of the 1920s and 1930s glamorized air-to-air fighting during World War I. The grim reality was that poor training killed many, and on some combat missions there was as much as a 70 percent chance of dying during the mission. Since parachutes were not used until the end of the war, fliers couldn't escape from doomed machines. This comic book cover illustration was by Frederick Blakeslee.

JUNKERS J 7

War accelerated aviation technology. Aircraft designer Hugo Junkers developed the all-metal, cantilever-wing J 1 monoplane to fill the need for a tough fighter aircraft. Later he would produce a similar ground-attack machine with armor plating for the crew. Junkers corrugated-metal machines were the shape of things to come.

cal advantage, experimented with new airplane technologies. Designer Hugo Junkers created the first all-metal military airplane, the Junkers J 1, whose cantilevered wings marked a major step toward the modern airplane. Another German innovator, Claudius Dornier, created the Dornier D I, an all-metal, semi-monocoque design that was the most advanced airplane produced during the war. While the Germans weren't ready to put such exotic craft in service, even their more conventional planes broke

new ground, thanks to Reinhold Platz, chief designer for Dutchman Anthony Fokker. The Fokker D. VII, whose maneuverability and excellent handling characteristics made it one of the best fighters in existence, was such an efficient killing machine that it was expressly specified in the Armistice agreement that all D. VII aircraft were to be surrendered.

Throughout the war, the demands placed on aircraft manufacturers for greater performance caused major im-

(BELOW AND BELOW, LEFT) **VOISIN VIII COCKPIT**

The photograph below shows the unrestored Voisin VIII cockpit. Visible are one of the rudder pedals, the aileron control wheel, and bomb sight window. Unlike most early bombers, the Voisin had an internal bomb bay; and the pilot could use a lever to select how many bombs to drop. Below, left, is a side view of the cockpit.

(ABOVE) **VOISIN VIII**

Delivered without an engine to the Smithsonian in 1918, this Voisin VIII was originally flown as a night-bombing demonstrator by the U.S. War Department. Restoration began in 1989 after the Museum obtained a motor from the U.S. Air Force Museum in Dayton, Ohio.

(OPPOSITE) **VOISIN VIII AT THE MUSEUM**

The National Air and Space Museum's Voisin VIII represents one of the most advanced military aircraft available early in the war. First used for night bombing in Germany in 1916, the rugged Voisin operated from Russian steppes to the Middle East. This machine is fitted with a Hotchkiss machine gun, but some Voisins carried 37mm cannons.

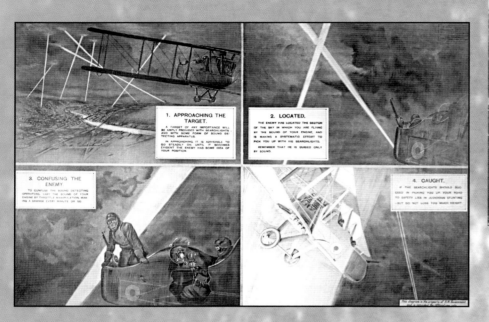

(ABOVE, LEFT) **TRAINING POSTER, "NIGHT FLYING, SEARCHLIGHTS"**

The British Royal Flying Corps tried to reduce casualty rates among novice aviators with a series of training posters. This one describes proper night-flying techniques.

(ABOVE, RIGHT) **AERIAL RECONNAISSANCE PHOTO**

The aerial photograph of Fort De Tavannes in the background—taken in September 1918—demonstrates how important aircraft had become to tactical planning on the ground.

(RIGHT) **TRAINING POSTER, "OUTMANEUVERED"**

An interesting top-view diagram shows new fighter pilots how to attack a German two-seater from below without being killed by the rear gunner.

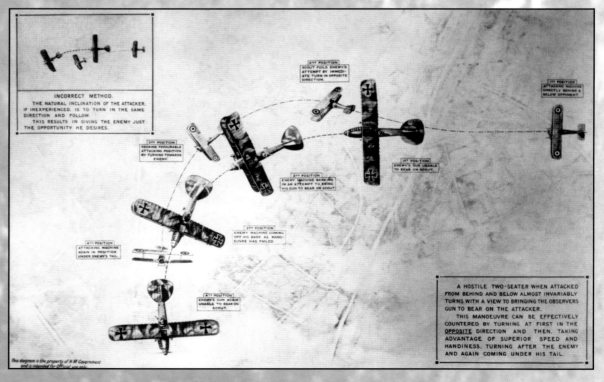

(ABOVE) DAYTON-WRIGHT DH-4 BOMBER AT THE MUSEUM

The National Air and Space Museum's Dayton-Wright DH-4 represents America's last-minute attempt to supply U.S. pilots with a combat aircraft. A British design, the DH-4 copy was obsolete when it entered service in 1918, and only a handful of squadrons flew them in combat.

(RIGHT) TRAINING POSTER, "BEWARE OF THE HUN IN THE SUN"

Another Royal Flying Corps training poster warns observers and gunners of a common German fighter pilot tactic. Attacked by a pair of machines, the backseater may not see the danger lurking in the sun.

(LEFT) **F.E.8 REPLICA AT THE MUSEUM**

The Air and Space Museum's *Legend, Memory, and the Great War in the Air* exhibition features this F.E.8 pusher fighter cutaway replica. Production delays made the F.E.8 dangerously obsolete by the time it entered combat in 1916. The Le Rhône 9J rotary engine was liable to crush the pilot during a crash.

(OPPOSITE) **SOPWITH CAMEL**

British pioneer aviator Thomas Sopwith produced the most successful fighter of the Great War in 1917: the Sopwith F.1 Camel. Responsible for destroying more than 1,290 enemy aircraft, the Camel was nimble in combat because most of its weight was concentrated near its center of gravity. Gyroscopic forces from the spinning rotary engine made for tight turns and difficult takeoffs and landings. Shown is a T.F.1.

Exclusive Post Card Edition   Active Service, No. 25

FRENCH AEROPLANE ATTACKING GERMAN AIRSHIP
The Airship is turning to retreat and is consequently foreshortened. The Aeroplane being so much nearer the Camera makes the Airship look small in proportion. This is the first photograph taken in mid-air of an aeroplane attack

£100 PRIZE WINNER

(RIGHT) **FRENCH AIRPLANE ATTACKING GERMAN AIRSHIP**

This photograph purports to show a French fighter, possibly a Farman, attacking a German airship. The terror of strategically bombing urban areas was pioneered by zeppelins. The giant dirigibles were never terribly successful, however—vulnerable to antiaircraft fire, their hydrogen gas envelopes would explode if hit with incendiary ammunition.

start of World War I, the Germans had adopted a program of strategic bombing, largely because they alone possessed a fleet of giant rigid airships built by Zeppelin and by Schülte-lanz, popularly called zeppelins. The zeppelins, however, were not well suited to the task; not only were the huge airships frequently blown off course, but they were too vulnerable to bad weather and the incendiary

machine gun fire of British and French pilots. By the spring of 1917 Germany had come up with a more formidable weapon— the giant twin-engine Gotha bomber. In June Gothas appeared over London—not in darkness, as the zeppelins had done,

but in broad daylight. At first the British defenses, organized for the zeppelins' nighttime raids, were caught unprepared. And while the Gothas could not carry as many bombs as the airships, the bombers' speed gave them an edge against enemy fighters. It wasn't long, however, before British guns were shooting down the attackers, forcing the Germans to operate at night once again.

ALBATROS D.VA

One of two left in the world, the Museum's Albatros D.Va sports the colorful markings that characterize German fighters of the period. The D.Va was easy to handle; such notables as Richthofen, Göring, and Udet flew them. This aircraft fought with *Jasta* 46 in the spring 1918 offensive before a bullet—still in its engine—brought it down.

(ABOVE) ***ZEPPELIN STAAKEN R.IV***

Riesenflugzeuge, giant airplanes, were Germany's answer to the failure of the airship bombing campaign. The Zeppelin Staaken R. IV had a wingspan of 138 feet—only three feet less than the wingspan of the B-29 Super-fortress of World War II. Inspired by Igor Sikorsky's *Il'ya Muromets,* the R-planes made their first night raid on Britain in September 1917.

(RIGHT) ***ZEPPELIN STAAKEN R.IV CREW IN FLIGHT***

Part of the crew of an R-plane in flight. Some of these behemoths carried seven crewmen, and eighteen 220-pound bombs. Twenty-eight sorties against England resulted in no casualties among the Zeppelin Staakens. Effective interception of night bombers by fighters would have to wait for radio, radar, and another war.

In September, London was bombed by the even more menacing R-planes, or *Riesenflugzeuge* (giant airplanes). Inspired by the success of Russia's *Il'ya Muromets*, these multi-engine giants were epitomized by the Zeppelin Staaken R. IV, R. V, and R. VI, with wingspans of 138 feet and the ability to drop bombs as large as 2,200 pounds. But even the R-planes did not reach the limit of the Germans' wartime dreams. There is evidence that one company was trying to build an even bigger craft. The Poll giant triplane, whose wings would have stretched almost thirty feet longer than the R. IV's, was envisioned as a transatlantic warplane capable of reaching the United States. Whether it could have even gotten off the ground will never be known; a lack of funds halted construction before the end of the war.

In any event, the physical damage wrought by German bombs was outweighed by their psychological impact: the bombs that fell on London shattered the distinction between soldier and civilian. Initially, it was the poor accuracy of the zeppelins that led the Kaiser to permit this breach of military convention. By the time of the Gotha and Staaken raids, a new rationale had been constructed in which the raids were designed to undermine the enemy's will to fight. And while that did not occur, the bombing raids of World War I showed that the airplane had become the instrument of total war.

Even before the war's end, the airplane's deadly new role made its impact. In Britain, where Londoners demanded retaliation for German raids and better civil defense, the Royal Air Force was created on April 1, 1918. As the first true air force, the RAF, with its twin responsibilities—to defend the British Isles and carry out offensive missions—set an example that other nations would follow. In time, the world's air forces would become institutions apart from, and competing with, the other military services. When the Armistice was signed in November 1918, the airplane's military potential was no longer in doubt. And the crucible of war had hastened a maturing process that would continue into the next decade, when the airplane would begin to shrink the globe. ■

(LEFT) COMMEMORATIVE BUST OF EUGENE BULLARD

Fleeing racism in his own country, American Eugene Jacques Bullard settled in France and joined the Foreign Legion when war broke out. He was twice wounded and declared disabled, but he volunteered for pilot training. Even after twenty combat missions, he was barred from flying with the U.S. Army Air Service.

(BELOW) GERMAN PILOT DEMONSTRATES OXYGEN EQUIPMENT

This Fokker D. VII pilot models late-war pilot's gear. Aircraft capable of high-altitude flight brought new hazards—above 10,000 feet, airmen required warm clothes and, more important, a supply of oxygen to stave off hypoxia. Only German pilots had the benefit of oxygen equipment.

(RIGHT) **RAY BROOKS WITH SPAD XIII**

"The SPAD XIII fighter of World War One had all that I could wish for in combat," said American ace A. Raymond Brooks, photographed with his French mount in 1918. Brooks scored one victory with 139th Aero Squadron before becoming a flight commander with the 22nd Aero Squadron in August.

(BELOW) **SPAD XIII** *SMITH IV* **AT THE MUSEUM**

Brooks's SPAD XIII *Smith IV* (named after Smith College, where his fiancée was a student) was restored by the National Air and Space Museum in 1986. Flown by Americans, French, British, Italians, and Russians, the rugged SPAD owed much of its success to its 220-horsepower, V-8 Hispano-Suiza engine. The shooting-star emblem of the 22nd Aero Squadron was designed by Brooks.

(ABOVE) DORNIER D I

Ferdinand Zeppelin's visionary partner Claudius Dornier created the all-metal, stressed-skin D I fighter for the single-seat fighter competition of June 1918. Even after killing pilot Wilhelm Reinhard in trials, the D I was judged by all the veteran pilots to be one of the best. Manufacturing flaws and a limited understanding of flutter and other aerodynamic problems plagued new aircraft designs.

(LEFT) DORNIER D I INTERIOR

This interior photo of the Dornier D I's fuselage shows how far aviation had come in five years of war. Dornier would go on to produce high-speed bombers for Hitler's Luftwaffe, but material shortages in Germany during 1918 prevented his and Junkers' metal aircraft from being produced in quantity.

(LEFT) FOKKER D. VII

Specifically banned in the Armistice that ended World War I, the Fokker D. VII was Germany's best all-around fighter. With it, the kill rate among the *Jastas* (German fighter squadrons) went from 217 in April to 565 in August. It wasn't fast, but its steep rate of climb made it lethal from below.

(BELOW) FOKKER D. VII AT THE MUSEUM

The National Air and Space Museum's Fokker D. VII as photographed in the Legend gallery. Incredibly, this machine was captured by three American airmen when it landed accidentally on an Allied frontline airfield two days before the Armistice.

(LEFT) **PFALZ D.XII CONSTRUCTION**

Factory workers construct a semi-monocoque plywood fuselage for the Fokker D.VII's stablemate, the Pfalz D.XII. Built as a replacement for aging Pfalz D.IIIs and Fokker triplanes, 800 D.XIIs somehow reached the front before the end of the war.

PFALZ D.XII AT THE MUSEUM

The Air and Space Museum's Pfalz D.XII sports the colorful but historically inaccurate paint scheme it wore in the 1930 Hollywood classic *Dawn Patrol*. The history of this Pfalz D.XII is better documented in its Hollywood career than in its wartime service.

PFALZ D.XII CONSTRUCTION

(LEFT) Factory workers construct a semi-monocoque plywood fuselage for the Fokker D.VII's stablemate, the Pfalz D.XII. Built as a replacement for aging Pfalz D.IIIs and Fokker triplanes, 800 D.XIIs somehow reached the front before the end of the war.

PFALZ D.XII AT THE MUSEUM

The Air and Space Museum's Pfalz D.XII sports the colorful but historically inaccurate paint scheme it wore in the 1930 Hollywood classic *Dawn Patrol*. The history of this Pfalz D.XII is better documented in its Hollywood career than in its wartime service.

(LEFT) RECRUITING POSTER

Propaganda poster for the U.S. Army Air Service reveals high hopes for America's entry into World War I. The country that invented the airplane had only several dozen absolescent military aircraft in April 1917, and U.S. pilots would fly French and British machines once they reached Europe.

(BELOW) WILLIAM MITCHELL

Alone among U.S. Army officers, William "Billy" Mitchell advocated a separate powerful air force. In September 1918 Brigadier General Mitchell commanded an air armada of 1,500 French and American aircraft, employing up to 200 at a time for bombing raids in the Meuse-Argonne sector.

CHAPTER FOUR

THE AIRPLANE COMES OF AGE

THE RYAN NYP (NEW YORK TO PARIS)

The *Spirit of St. Louis,* in which unknown airmail pilot Charles Lindbergh made the first solo flight across the Atlantic Ocean in May 1927. Lindbergh guided his tiny monoplane into history—3,610 miles in thirty-three hours. At left, a close-up of the Ryan's reliable Wright J-5 Whirlwind engine. Above, Lindbergh was photographed in front of his mount—note the lack of forward-looking windows.

Early in the wet, gray morning of May 20, 1927, twelve-year-old Isabel Haynes and her family stood in a small crowd of onlookers at Long Island's Roosevelt Field. Before them, a lanky airmail pilot with a serious and boyish face was about to take off on a flight that would either bring him, in his words, "death and failure" or make him the most celebrated American of the twentieth century. Isabel had heard of Charles Lindbergh's quest to make the first nonstop flight across the Atlantic, but only now did she see the man the press had lately been so excited about. Decades later she would remember her own excitement as Lindbergh, standing next to the Ryan monoplane he had christened the *Spirit of St. Louis,* deliberated with his assistants about whether to go. Drizzle was falling at Roosevelt Field, but a high-pressure system was sweeping the rain out to sea. Lindbergh would follow.

In recent months a host of fliers had challenged the Atlantic and failed; six had died and three others had been seriously injured. Now this all-but-unknown pilot would stake his life on yet another attempt and, unlike the others, he would fly alone. In Paris, in London, and even here at home, many called his plan crazy. This despite Lindbergh's months of preparation, helping to design his plane and oversee its construction, meticulously plotting his course to Paris. Nothing was haphazard about this flight or about the seasoned and willful pilot who had already risked his life many times in the air.

At last Lindbergh was ready, and at his signal the engine rumbled to life. Helpers grabbed the wings and began to push the plane, which was overloaded with fuel, down the muddy field. They dropped back as Lindbergh picked up speed. Finally, the *Spirit* bounced into the air and stayed there, straining to ascend, as the crowd ran after it. Everyone looked anxiously at the telephone wires just beyond the far end of the field, hoping Lindbergh would clear them. When he did, by about twenty feet, the crowd cheered.

For the next thirty-three hours Isabel followed the flight on the radio, staying up as late as her parents would let her, awakening at the crack of dawn the next morning. She could not know of Lindbergh's sleepless ordeal, of his battles with rain, ice, and exhaustion, as he flew onward, pushing the limits of his endurance. She only knew, with a child's certitude, that he would make it.

When he did just that, struggling

SPIRIT OF ST. LOUIS AT THE MUSEUM

The *Spirit of St. Louis* now hangs in the National Air and Space Museum's Milestones of Flight hall. The NYP, less stable than the Ryan M-2 from which it was developed, did not allow Lindbergh to relax much during the flight. Flags on the aircraft's nose were added during Lindbergh's post-flight tour of Latin America and the Caribbean. (left) The instrument panel of the *Spirit* featured a T-shaped inclinometer to monitor altitude; below the panel is a maze of fuel pipes and valves.

AFTER THE FLIGHT

Charles Lindbergh waves to cheering Parisians from the town hall window. As they did with Wilbur Wright before him, the French took Lindbergh to heart, despite the disappearance of war hero Charles Nungesser and François Coli after leaving France two weeks before. (below) Lindbergh later came home, by U.S. Navy ship, to a hero's welcome—a ticker-tape parade in New York City, where 4 million people lined the route. (below, right) The ornate $25,000 check Lindbergh received for winning the Orteig prize.

New York June 17th, 1927

Bryant Park Bank

Pay to the Order of Charles A. Lindbergh

Twenty-five Thousand No/100 Dollars

Payable in funds current at New York Clearing House

$25,000. No/100

Raymond Orteig

through fatigue to bring the *Spirit* down safely at Le Bourget field, Lindbergh was greeted by an ecstatic throng of Parisians shouting his name. The flier who would now be called the "Lone Eagle" had risen to a status no aviator had dreamed of. The powerful of Europe opened their doors to him. Everywhere he went, he was mobbed by admirers. He navigated his sudden fame with a poise that only heightened their admiration. And when he returned home to parades and state dinners, he was more than a hero; he was the embodiment of America's faith in its own dreams, however ambitious.

Lindbergh himself had dreams for the future of aviation, and he was determined to use his fame to champion the cause with the American public. The celebrations had hardly died down when he took off in the *Spirit* on a nationwide tour, stopping in all forty-eight states. His message was a simple one: from now on, America's progress would be measured, most of all, in the sky. Cities that built airfields would prosper ahead of those that did not. And by logging

NC-4 CROSSING THE ATLANTIC

Lindbergh was not the first to fly across the Atlantic; approximately 100 had preceded him. The crew of the U.S. Navy's NC-4 flying boat were the first, as seen in this painting by Ted Wilbur. This May 1919 feat proved the reliability of the big Curtiss seaplane—even as two other accompanying NCs failed in the attempt.

(ABOVE) **BESSIE COLEMAN**

The first African-American woman aviator, Bessie Coleman received her pilot's certificate in 1921 in France. Like many aviators in this era, Coleman flew an old Curtiss JN-4 "Jenny" World War I trainer aircraft for her shows. She was killed in 1926 during a practice parachute jump.

(LEFT) **BARNSTORMING**

Brave souls—including Charles Lindbergh—toured the United States after World War I, thrilling small town America with their "barnstorming" exploits. Here, Mabel Cody transfers from a water ski to a war-surplus trainer. Cody was the first woman to transfer from a speedboat to an airplane, off the coast of Davis Island, Florida, in May 1927.

more than 22,000 air miles in three months, making eighty-two stops without an accident and, with one exception, always on schedule, Lindbergh showed that flying had become safe and reliable. The tour helped spur Americans to embrace aviation as they never had before in the quarter century since the Wright brothers flew. But in 1927 a nonstop flight from New York to Paris was just the most visible in a spate of changes that had already begun to transform U.S. aviation. Within a decade Americans would make their own transoceanic journeys in the first truly modern airliners. The United States was on its way to becoming the world's preeminent nation in the air.

The United States had entered the 1920s with airplane production at only a fraction of the prolific levels reached by the end of World War I. And though the war had clearly defined military roles for the airplane, aviation had not yet found a niche in civilian life, aside from the barnstormers and stunt pilots who barely made a living with their exploits. Unfortunately, the barnstormers only managed to reinforce the public's belief that airplanes were dangerous, a perception that hindered the handful of entrepreneurs struggling to develop airlines. But in Europe, thanks mostly to government subsidies, passenger aviation was thriving. Germany, whose economy had been destroyed by the war,

(LEFT) AIR MAIL PILOT BILL HOPSON

William "Wild Bill" Hopson epitomizes the unsung heroes of the U.S. Airmail Service. Outdated equipment, poor weather, and primitive fields didn't prevent them from moving 49 million letters by 1920 or spanning the continent in less than twenty-six hours early in 1921. Hopson died in a crash before the decade was over.

(BELOW) EARLY COMMERCIAL AVIATION

Reliable aircraft made aviation more practical in the 1920s, but no less dangerous: forty pilots were hired by the U.S. Post Office Department to fly the mail in 1918; twelve of them had been killed by the end of 1920. The airplane shown is a Western Air Express Douglas M-2, contracted to fly the mail. After 1926, contract airmail carriers, precursors of the modern airlines, carried the mail. The background photograph is an early example of cropdusting as a Curtiss JN-4 sprays a field in Ohio, 1921.

(ABOVE) DOUGLAS WORLD CRUISER—*CHICAGO*

Safely ensconced in the National Air and Space Museum, the Douglas World Cruiser *Chicago* flew a punishing 26,345 miles in six months, circumnavigating the globe in 1924. *Chicago* and her sister ship *New Orleans* did their part to use up war-surplus Liberty engines—nine apiece—in their record-setting flight. The round-the-world flight was a scheme by the U.S. Army Air Service to prove that the airplane was a viable method of transportation. Four aircraft named *Seattle, Chicago, Boston,* and *New Orleans* took off from Seattle on April 6, 1924. Both *Chicago* and *New Orleans* finished the trip.

(LEFT) *SEATTLE* IN ALASKA

U.S. Army Air Service World Cruiser *Seattle* was photographed at Kanatak, Alaska, shortly before crashing early in the round-the-world attempt. The equally ill-fated *Boston* was forced down in the North Atlantic near the end of the flight. The around-the-world flight was the brainchild of William "Billy" Mitchell.

93

Wright Whirlwind engines (the same radial, air-cooled design used in the *Spirit of St. Louis*) or, in later versions, the more powerful Pratt & Whitney Wasp, the Ford's cruising speed was 105 miles per hour, enough to manage a transcontinental trip in two days. The Tin Goose had its drawbacks: within the corrugated-metal cabin walls, the noise of the engines was so bad that passengers were given earplugs for the journey. The NACA engineers worked systematically to identify the causes of drag and were able to show why the Tri-Motor, like most of its contemporaries, was an aerodynamic underachiever. The Ford's slab-sided fuselage was partly to blame, but even worse was its fixed landing gear and underslung engines. Everyone had known that fixed gear contributed to an airplane's drag, but no one knew how much—until NACA wind tunnel tests pegged it at a whopping 40 percent.

Another culprit, the air-cooled engine, with its exposed cylinders, was a challenge made for NACA researchers. Using the Propeller Research Tunnel, they were startled to discover that such engines accounted for nearly one-third of the total drag from an aircraft's fuselage. There was no simple solution, because covering the engine would prevent it from being cooled by the airstream. Or would it? At the NACA, the surprising discovery was made that a specially shaped streamlined cowling would actually promote better cooling.

This was such a breakthrough that in 1929 its inventors won the prestigious Collier trophy, the premier award for aeronautical advancements.

The NACA cowling, as it came to be known, proved itself on such airplanes as the six-seat Lockheed Vega, whose monocoque fuselage was reminiscent of France's prewar Deperdussin racer. But whereas the racer had required painstaking labor to build, Lockheed's brilliant designer Jack Northrop had devised a means of prefabricating the Vega's wooden shells, greatly simplifying construction. The Vega also incorporated cantilevered (that is, internally braced) wings, pioneered over a decade earlier by German designer Hugo Junkers. Together, these features reduced drag so successfully that by 1928 the 5C Vega was the world's fastest airliner, cruising at 160 miles per hour. Even that wasn't good enough for Lockheed designers, who lowered the Vega's wing and incorporated retractable landing gear to produce the Orion, with a cruising speed of 200 miles per hour and a top speed of 226 miles per hour. Suddenly, a commercial airliner could outrun almost any fighter plane in existence.

But for the airlines, speed alone was not enough; it was essential to build larger planes with increased passenger capacity and the safety of multiple engines. That need forced designers to confront their greatest obstacle: weight. Ever since the

(OPPOSITE, TOP) **NORTHROP GAMMA *POLAR STAR***

Fuselage wrinkled from a hard landing, the Northrop Gamma 2B *Polar Star* flew 2,400 miles across Antarctica in 1935. En route explorer Lincoln Ellsworth and pilot Herbert Hollick-Kenyon spent an entire day scooping snow out of this machine with a tea cup following a storm.

(OPPOSITE, BOTTOM) ***TINGMISSARTOQ* AT THE MUSEUM**

Charles Lindbergh's Lockheed Sirius *Tingmissartoq* benefited from NACA research in the form of a streamlined cowl for its radial engine. With this airplane, Lindbergh and his wife, Anne, explored flight routes across both the Pacific and Atlantic Oceans. The plane was christened *Tingmissartoq* ("the man who flies like a big bird") by an Eskimo boy in Greenland. The NACA kept the United States in the forefront of aviation research as war clouds once again gathered in Europe.

(OPPOSITE) **AKRON AFLOAT**

Materializing out of the clouds, the Navy airship USS *Akron* flies over Maxwell Field, Alabama, in 1932. A storm destroyed the giant at sea, killing seventy-three officers and men and Admiral Willian Moffett, the head of Naval Aviation. *Akron*'s sister ship USS *Macon* was lost in a similar accident.

THE AIRSHIP'S GOLDEN ERA

If the period between the wars was a formative one for the airplane, it was a golden era for the airship. Used as weapons in World War I, these airborne giants offered the chance to fly across the Atlantic in amazing quiet and often luxurious comfort. Several nations experimented with airships, but only Germany came close to mastering them; its *Graf Zeppelin* and, later, the giant *Hindenburg* became icons. But the airships were doomed from the start. Slow, unwieldy, and difficult to maneuver, they were also sensitive to changes in air currents, making them ill-suited to flights in questionable weather. And as the U.S. Navy discovered, they were dangerous: three of its airships—the *Shenandoah*, the *Akron*, and the *Macon*—met tragic ends.

(ABOVE) **GERMAN AIRSHIPS**

Two icons of Germany's airship era, the Graf Zeppelin and the Hindenburg, are featured in this contemporary advertisement for motor oil

(RIGHT) **NAVY ZRS-4 *AKRON* CHRISTENING**

The USS *Akron* and *Macon,* built in the United States by Goodyear-Zeppelin Company, were important parts of the Navy's defense strategy. Both used helium for lift and actually carried scout aircraft launched from a trapeze-like mounting.

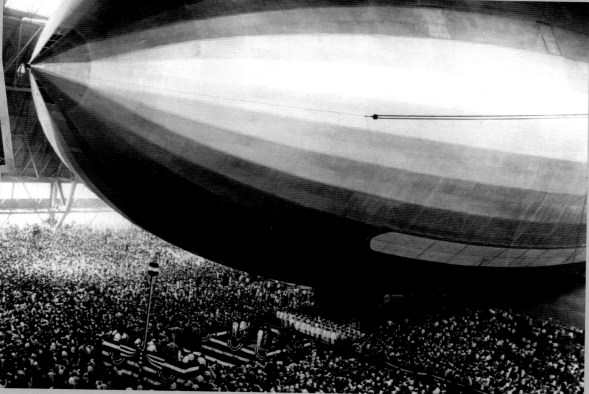

(OPPOSITE) AIR TRANSPORTATION GALLERY

Twenty years of commercial aviation grace the Air and Space Museum's Air Transportation hall. Counterclockwise from upper left: the Douglas DC-3, the Fairchild FC-2, Boeing's 247D, the Grumman G-21 "Goose" amphibian, Northrop's Alpha passenger plane, and the Pitcairn Mailwing airmail plane.

(ABOVE) REACHING FARTHER: DC-2 IN SOUTH AMERICA

Llamas helped promote Pan American–Grace Airways (Panagra) in Peru in the 1930s. Douglas designed the DC-2 aircraft seen in the background to compete with Boeing's 247D, which was operating with United Air Lines. American Airlines, however, needed something bigger and turned to Douglas for help.

airplane could carry people and turn a profit.

The DC-3 eclipsed all airliners that came before it, including the DC-2, and became a workhorse of legendary status. And it came just in time for the airlines, which had been thrown into turmoil following the airmail cancellation crisis and the passage of the Air Mail Act of 1934. The carriers' prosperity in the midst of the Depression had made them a target for political scrutiny, and with the administration of Franklin Roosevelt came a crackdown on perceived improprieties in the airline business—specifically, the giant holding companies that served as parents of both airlines and aircraft manufacturers. The new Air Mail Act broke up the holding companies and greatly reduced airmail subsidies to the airlines. For four years, until the situation was ameliorated, the DC-3 allowed the struggling airlines to remain solvent. And after the creation of the Civil Aeronautics Authority in 1938, the prospects further improved with centralized, rational regulation to help the airlines grow and prosper.

Even as the DC-3's capable presence appeared around the world, a parallel revolution was under way on the seas, where flying boats blazed new trails in transoceanic travel. These aircraft, with their ability to take off and land on any clear stretch of water, fulfilled Igor Sikorsky's vision for large-capacity, comfort-

THE AIRPLANE COMES OF AGE

(RIGHT) SPANNING THE PACIFIC

Russian designer Igor Sikorsky brought his talent for multi-engine aircraft to America. He test-flew his S-42 flying boat in 1934. That year it set ten seaplane records and scouted Pacific routes for Pan Am, opening the way for later passenger service using such planes as the Martin M-130.

(BELOW) SIKORSKY S-42 INTERIOR

For passenger comfort, the large flying boats of the 1930s were in a class by themselves. Most of the world still lacked airports, and the oceans offered large runways.

(LEFT) **JUAN TRIPPE WITH CHARLES LINDBERGH**

Charles Lindbergh was hired by Pan American Airways president Juan Trippe to pioneer airline routes for the company. He had spent two months visiting every South and Central American country bordering the Caribbean and the Gulf of Mexico in 1927.

(BELOW) **SIKORSKY S-43 AT THE MUSEUM'S GARBER FACILITY**

The Air and Space Museum's S-43 (JRS-1) patiently awaits restoration at the Garber storage facility. The boat-shaped fuselage recalls Curtiss designs, and the retractable wheels allowed the S-43 to use boat ramps.

(LEFT) **AFTER PLOESTI**

Members of the 93rd Bomb Group in Benghazi, Libya, pose after their harrowing August 1, 1943, raid on the oil refinery in Ploesti, Romania. Major Ramsay D. Potts, Jr., who piloted one of the B-24 Liberators that carried out the raid, is seated fifth from left.

(OPPOSITE) **MACCHI C.202 FOLGORE AT THE MUSEUM**

The Macchi Company designed a fuselage around Germany's Daimler-Benz liquid-cooled in-line engine in 1940. The resulting C.202 *Folgore* (lightning) was an Italian fighter pilot's dream—with an extra eight and three-eighths inches of left wing to counteract engine torque. The North American P-51 Mustang, one of the best fighters of World War II, is shown below the Macchi.

the bomber as the ultimate weapon. The key, they said, was that aerial bombardment would bring the battle to the enemy's heartland. Many listeners, realizing that innocent civilians would face as much danger as soldiers on the front lines did, were horrified. But their dire predictions only strengthened the advocates' case. British Prime Minister Stanley Baldwin, addressing Parliament on the eve of the 1932 Geneva Disarmament Conference, warned that "no power on Earth" would protect citizens from a rain of destruction. "The bomber," said Baldwin, "will always get through." As World War II opened, the air forces of England and Germany alike believed that such attacks would ensure

victory by shattering the enemy's will to fight. In the United States another concept took hold: precision bombing, in which surgical air strikes at carefully selected targets would cripple the enemy's capacity to wage war.

But as Ramsay Potts knew well, the bomber didn't always get through. Squadrons based in England that tried to strike targets inside Germany did so without fighters to defend them, because no existing fighter had enough range to make the round trip. The attacking planes found themselves vulnerable to German fighters and antiaircraft guns, often with tragic results.

But Ploesti, the 93rd hoped, would be

P-47 PILOT QUENTIN AANENSON

Quentin Aanenson flew P-47 Thunderbolts in the 391st Fighter Squadron. Veteran of numerous missions to Germany and France, Aanenson was haunted by the destruction he could inflict with this big fighter-bomber.

NOVEMBER 1944

"When the personal stories of this war are told, the world will find it all difficult to believe. Without actually being under fire, no one can possibly understand what hell it can be." It had been months since Quentin Aanenson had written these words to his fiancée, Jackie; and in that time he had amassed enough painful memories to last a lifetime. As a twenty-three-year-old Thunderbolt pilot with the 391st Fighter Squadron, Aanenson had entered the war shortly before D day and had flown dozens of strafing and bombing missions over France and Germany. And he had seen his share of miracles.

Many times, sent behind enemy lines to attack German air bases, he had steered his plane through walls of flak; somehow he had avoided being killed. In August 1944 his P-47 had been hit by antiaircraft shells and burst into flames. Trapped in the burning cockpit by a jammed canopy, Aanenson had faced a pilot's worst nightmare. He put the Thunderbolt into a steep dive, intending to hit the ground as fast as possible. But his suicide attempt saved his life: the rush of air sucked the flame out of the cockpit and the fire died out. Then in September his aircraft was nearly blown in two by an 88mm shell. The Thunderbolt held together long enough for him to reach his airfield, but on landing, the plane's tail section collapsed, the two parts held together by a set of control cables as Aanenson rolled down the runway. When the other pilots saw the wreckage, they were amazed that he had survived.

Many of his squadron mates hadn't been so lucky. It was hard enough to lose friends in combat, but Aanenson was just as distressed to see men die by his own hand. He would never forget the sight of his P-47's machine guns tearing into German soldiers, some of whom were blown backward by the force of the bullets' impact. In the moment, he had thought only of doing the job he had been trained for. But after he had returned to base, the memory of it made him sick. Most of the time, though, he never saw the men he killed. By the fall of 1944, as the war dragged on, Aanenson had to admit he had become hardened to the horrors of this war; but he never got used to the killing.

Now, on a late November day, Aanenson and his wingman were strafing enemy positions on the edge of a small German village when he heard on his radio a frantic plea for help from an infantry captain below. He and his men were about to be attacked by a German tank, which was heading straight for them. Aanenson could see it, but he told the captain he was afraid his 500-pound bombs might miss the tank and kill the Americans. The infantryman told Aanenson that if the tank weren't stopped, they'd be dead anyway. Hearing this, Aanenson told the captain to disperse his men, that he was coming in right away.

B-17 MURAL, *FORTRESS UNDER FIRE*

Artist Keith Ferris froze time with his mural of an actual 303rd Bomb Group mission over Wiesbaden, Germany, in 1944.
Thirteen 50-caliber machine guns meant the Flying Fortress lived up to its name. This is Ferris's sketch for the final mural,
which hangs at the Museum.

BATTLE-DAMAGED B-17

Exploding flak during a raid over Hungary killed both waist gunners and crippled this Fifteenth Air Force Flying Fortress, which amazingly limped 600 miles to a safe landing.

(ABOVE) *FLAK BAIT* **WITH CREW**

The appropriately named Martin B-26 Marauder *Flak Bait* has more than 1,000 shell-hole patches from 725 hours of combat. With 202 missions to Germany, France, Belgium, and Holland, this veteran bird retired to the Air and Space Museum.

(LEFT) **B-26 MARAUDER INTERIOR**

Flak Bait's radio and navigation stations appear in this photograph. Just visible through the center door is the cockpit instrument panel, which was shattered on September 10, 1943, when a 20mm shell from a Bf-109 hit it from behind. The wounded pilot later made a perfect landing.

FLAK BAIT AT THE MUSEUM

The nose section of *Flak Bait*, seen in the Museum's World War II Aviation gallery, still bears the scars of battle. Recalled *Flak Bait*'s pilot, James J. Farrell, "It was hit plenty of times, hit *all* the time." Patches on the aircraft's skin cover bullet holes, proving that *Flak Bait* deserved its name.

A TIME OF HEROES

(RIGHT) MITSUBISHI ZERO IN FLIGHT

A rude shock to Western military analysts, Mitsubishi's Zero fighter had excellent range and gave veteran Japanese airmen a decided advantage. At 6,025 pounds loaded, the Zero sacrificed pilot armor and self-sealing fuel tanks to save weight for maneuverability.

(BELOW) MITSUBISHI ZERO AT THE MUSEUM

The National Air and Space Museum's A6M5 *Rei-sen* (Zero fighter) model 52 was one of several aircraft captured on Saipan in 1944. It flew at Wright and Eglin Fields in the United States as part of flight evaluations.

As the Allied armies continued their long and determined advance on Berlin, the other war—the struggle against the aggression of the Japanese empire—was still raging in the Pacific. There it was already clear that naval warfare was no longer the domain of the battleship; now it belonged to the aircraft carrier. These floating airfields, their decks bristling with warplanes, were the most powerful weapon in the war at sea. In 1942 U.S. aircraft carriers had brought about the stunning defeat of Japanese forces at the island of Midway, breaking the enemy's hold on the Pacific. By the fall of 1944 the United States was about to fulfill the apprehension voiced by Japanese Admiral Isoroku Yamamoto when he learned the results of the Pearl Harbor attack: "I fear all we have done is awaken a slumbering giant and fill him with a terrible resolve." For U.S. carrier pilots, the airplane that epitomized that resolve was the dive bomber.

AIRCRAFT CARRIERS IN THE PACIFIC

Deck crowded with Hellcats, Avengers, and Helldivers, the U.S. Navy carrier *Essex* leads a task force in the Pacific. During the 1944 invasion of the Marianas, *Essex* and her sister carriers destroyed the last of Japan's naval air arm.

A TIME OF HEROES

(RIGHT) GLOSTER METEOR

The Gloster Meteor, Britain's first jet fighter, was the only Allied jet to see action in World War II. It began flying for the RAF in 1944 to help intercept German V-1 flying bombs.

(BELOW) ME 262: GERMAN JET FIGHTER AT THE PAUL E. GARBER FACILITY

A terrifying sight to Allied fighter pilots over Germany, Messerschmitt's Me 262 *Schwalbe* (swallow) exceeded the speed of propeller-driven fighters by as much as 100 miles per hour. Four 30mm cannons made the Me 262 a formidable adversary, but U.S. Army Air Force pilots learned to attack them on takeoff and landing to even the odds.

(BELOW, RIGHT) STRAFING AND COMBAT

As the caption for this 1944 photo collage reports: the Eighth Air Force fighter who caught up with this Messerschmitt Me 262A over Germany had all the instincts of a recognition expert, for as he fired a series of bursts, he photographed it from half a dozen angles, thus providing a group of pictures that gave the best idea at the time of the appearance of this new airplane.

THE Me-262

The Eighth Air Force fighter pilot who caught up with this twin-jet enemy aircraft over Germany had all the instincts of a recognition expert, for as he fired a series of bursts, he photographed it from half-a-dozen angles, thus providing a group of pictures which give the best idea to date of the appearance of this new plane.

55305 A-C

score against the Me 262. The secret was to wait until the German pilots ran low on fuel; then, during the jets' landing approaches, the Allies attacked them. Although the Me 262's cannons downed a number of enemy planes, Allied fliers shot down the jets in almost equal numbers.

But there was no way around Germany's dominance of jet aviation in World War II, although Britain's Gloster Meteor, which first flew in 1944, was built in great numbers. The first U.S. jet fighter, the P-59, never saw combat; the prototype XP-59A now hangs in the Milestones gallery.

In the 1950s, jets took their place as the most important aviation technology. In military flying, they became important weapons on both sides of the Korean War. In commercial aviation, they suffered setbacks with the failures of Britain's Comet jet airliner, but by 1958, jets were firmly established at the forefront of air transport.

(LEFT) V-1 TAKING OFF

German secret weapons continued to proliferate throughout the war. The Arado Ar 234 jet was converted to a nearly unstoppable bomber, and the dreaded V-1 flying bomb began landing on England in June 1944. Driven by a primitive pulse-jet, the V-1 was a high-priority target for Royal Air Force pilots, who used cannon fire and wing-tipping to stop them.

(ABOVE) THE FIRST JET BOMBER

The Arado Ar 234 *Blitz* (lightning), the first jet-powered bomber, flew a small number of missions toward the end of the war. It was also used for high-altitude photo-reconnaissance, with its ability to evade Allied fighters. This view shows the cockpit of the Museum's Ar 234.

171

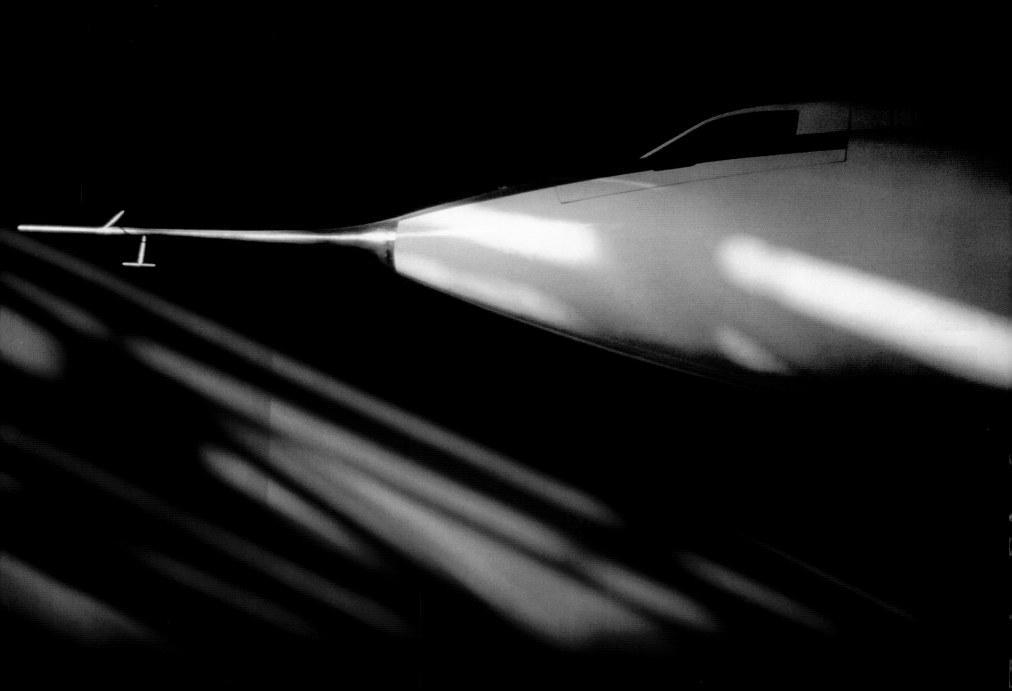

CHAPTER SIX

ROCKETS FOR THE COLD WAR

The "Prisoners of Peace" began arriving in the fall of 1945. That, at any rate, was what the German rocket scientists and engineers jokingly called themselves after beginning new lives at the U.S. Army's Fort Bliss in El Paso, Texas. Numbering more than a hundred, they were members of the team that had developed the Third Reich's V-2 ballistic missile. Within days after Germany's defeat they had sought out American GIs, offering to surrender. Now these men, led by a brilliant young engineer named Wernher von Braun, had been brought to the American Southwest—along with captured parts for several dozen V-2s—to help give a boost to U.S. rocket technology.

By this time American rocketry was two decades old, having gotten its start with a solitary dreamer named Robert Goddard. This quiet and reclusive physics professor from Worcester, Massachusetts, had become intrigued by the notion of space travel as a turn-of-the-century teenager.

(RIGHT) **GERMANY'S ROCKET PIONEER**

One of the architects of modern rocketry, Wernher von Braun is seen here (right) carrying a rocket model during his activities with the German Rocket Society about 1930. Von Braun went on to be technical director of Nazi Germany's missile-development program.

(OPPOSITE) **BUILT FOR SPEED**

First to fly double the speed of sound, the Douglas D-558-2 Skyrocket was one of several rocket-powered research aircraft created after World War II. On November 20, 1953, flying this Skyrocket in a shallow dive, test pilot Scott Crossfield reached a speed of Mach 2.005. The D-558-2 now hangs in the Milestones gallery.

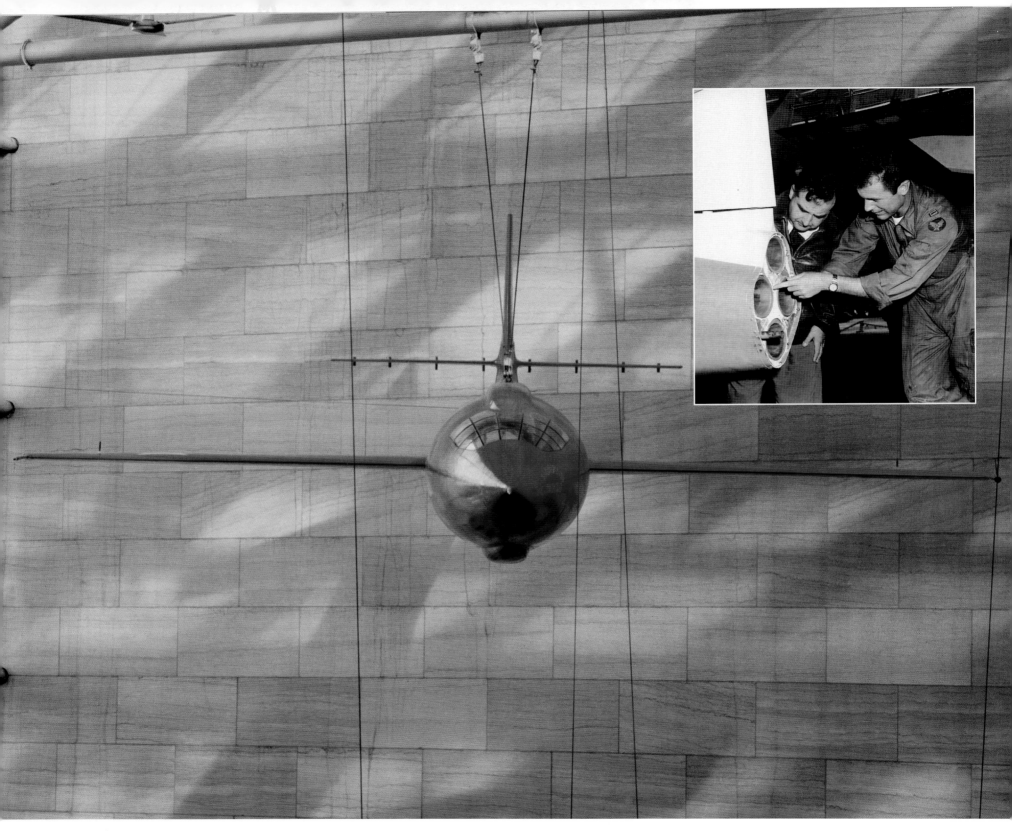

(OPPOSITE) X-1 AT THE MUSEUM

Glamorous Glennis hangs in the Milestones gallery. Its orange color was chosen to give the X-1 maximum visibility, for the benefit of tracking cameras and chase planes. (Later, white was found to be a more effective color.)

(OPPOSITE, INSET) THE X-1 COMES TO THE MUSEUM

Yeager (right) and Paul Garber inspect the thrust chambers of the X-1 after its arrival at Maryland's Andrews Air Force Base on August 28, 1950, en route to the National Air Museum, as it was then called. It was one of the rare times when both plane and pilot were on hand for the occasion.

(RIGHT) MX-774 TEST LAUNCH

The Air Force's MX-774, one of the earliest U.S. attempts to learn from the V-2, lifts off from New Mexico's White Sands Proving Ground in 1948.

sense of the potential for future conflicts, some farsighted military leaders, most notably General "Hap" Arnold, wartime commander of the Army Air Forces, looked at the V-2 and saw a stepping-stone to the ultimate weapon, a nuclear-tipped intercontinental ballistic missile (ICBM) that could be used against any enemy, even one halfway around the world. A few realized that the same technology could be used to boost a spacecraft into orbit, for military reconnaissance or for the peaceful exploration of space. Such was the double-

edged potential when the German rocket team arrived in America.

In the immediate postwar years, however, the prospects for rockets were mixed. There was enough demand for rocket technology in the defense industry to support the world's first rocket companies. These included Reaction Motors Inc. (RMI), which had been created shortly after Pearl Harbor by members of the American Rocket Society, and Aerojet, Inc., formed by a group of JPL engineers. In 1947 an airplane called the Bell X-1, powered by an

RMI rocket engine, propelled Air Force test pilot Chuck Yeager through the sound barrier on the world's first manned supersonic flight. In the years to come, rocket planes would lead in aviation's quest to go higher and faster.

An ICBM was another matter. It would take a rocket many times more powerful than the V-2 to hurl a nuclear bomb thousands of miles. For all the power of the V-2's engine, its design was too inefficient to be used to make larger motors. Then there was the monumental task of guiding the warhead to its target. The poor performance of the V-2's guidance system, designed to steer the missile from German-occupied Holland and France to cities in Britain and Belgium, had caused many of its failures. Hitting a target in the Soviet Union from the United States, a distance of perhaps 5,000 miles, was beyond existing technology. Thoughts of intercontinental ballistic missiles seemed far-fetched at best. Amid such skepticism, and facing drastic postwar cuts in military budgets, the newly created U.S. Air Force placed its hopes for nuclear delivery systems in the next generation of manned bombers that were still on industry drawing boards. In 1947 funding for a short-lived Air Force ICBM research effort was canceled.

At Fort Bliss the German rocket engineers found themselves with relatively little to do. One of their first assignments had

ROCKETS FOR THE COLD WAR

(RIGHT) SPACE VISIONARY

By the early 1950s Wernher von Braun was communicating his vision of space exploration to a nationwide audience. Behind him, von Braun's designs for a multistage passenger rocket and a wheel-shaped space station—lavish even by today's standards—are rendered by artist Chesley Bonestell.

(BELOW) MARS EXPEDITION

A Mars expedition, designed by von Braun and painted by Bonestell in 1953. Although the surface of Bonestell's Mars bears a striking resemblance to the real one, Viking landers would later show the martian sky as salmon-colored, not blue.

hatred of Soviet Communism—stressed the military potential of orbiting human observers who would be able, he said, to detect troop movements, planes being readied for takeoff on the deck of an aircraft carrier, and other preparations for war. One of von Braun's rivals, Milton Rosen (who was building a sounding rocket called Viking for the Naval Research Laboratory, or NRL), criticized him for selling a technological fantasy to the public. But the *Collier's* series became a seminal act of consciousness-raising; at a time when space flight seemed like science fiction, the authors portrayed it as a grand adventure on the verge of reality.

The realities of the Cold War, however, were far more pressing. In August 1953, eight days before the Redstone missile's first launch success, the Soviet Union announced that it had exploded a hydrogen bomb—less than a year after the first U.S. thermonuclear test and far sooner than anyone in the West expected. The news turned up the heat on U.S. missile development. The awesome destructive capability of thermonuclear weapons—almost a thousand times more powerful than the

REDSTONE LAUNCH

Getting a boost from the engine designed for Navaho, von Braun's Redstone missile lifts off from New Mexico's White Sands proving ground in the late 1950s.

JET AVIATION TAKES OFF

The decade of the 1950s saw jet aviation undergo a rapid advance. Jets made their first appearance in combat during the Korean War. Many of these planes (including the U.S. F-86 and the Soviet MiG-15) had swept wings, a design feature identified for high-speed flight by aerodynamic researchers during World War II, mostly in Germany. As with rocketry, the expertise of captured German engineers in jet technology benefited the United States and Soviet Union, whose military jets began to achieve supersonic speeds in the years following the Korean conflict. At first, jet fighters had functioned much as their propeller-driven predecessors had, designed mostly for one-on-one skirmishes with enemy fighters. By the end of the decade, however, their role had changed. A new generation of jets, armed with air-to-air missiles and equipped with onboard radar, were designed to intercept enemy bombers. With the advent of small nuclear weapons, jet fighter-bombers were added to the U.S. and Soviet Cold War arsenals.

In commercial aviation, the introduction of jets was marred by tragedy. The first jet airliner, Britain's Comet, suffered a series of fatal crashes later traced to a structural flaw. In addition, early turbojet engines of the type used in the Comet were highly inefficient; they gulped fuel. Although kerosene (the main ingredient of jet fuel) was far cheaper than the aviation gas used for piston aircraft engines, it was not until 1958, with the Boeing 707, that jet airliners became fully established. Yet to come were refinements that would eliminate excess noise and pollution (the engines of early 707s issued clouds of black smoke). But even now, with operating costs falling dramatically, it was clear that jet transport was here to stay.

F-104A STARFIGHTER AT THE MUSEUM

One of the so-called Century series of supersonic jet aircraft, the Lockheed F-104 Starfighter was the first U.S. interceptor capable of flying at sustained speeds above Mach 2 (twice the speed of sound). Its turbojet engine, which took up more than half the length of the fuselage, produced 14,800 pounds of thrust. The Museum's F-104A, which was flown by NASA between 1956 and 1975, hangs above the west escalator in front of the Planetarium.

NORTHROP YB-49 FLYING WING

Forty years ahead of its time, Northrop's B-49 flying wing was a bold concept when it appeared in 1950. Shown here is a variant called the YRB-49-A, which featured four jet engines mounted in the wings and two more in external pods. Designers hoped that by doing away with the fuselage, thus eliminating a major source of drag, the B-49 would be able to fly farther with heavier loads of armaments. Unfortunately, the plane was too unstable in pitch and yaw to be useful as a bomber. Not until the 1980s would computer control allow the flying wing to be used successfully (in the B-2).

bomb that destroyed Hiroshima—meant that an ICBM no longer needed pinpoint accuracy to be effective. Equally important, hydrogen bombs of equivalent destructive power weighed substantially less than Hiroshima-type atomic weapons. Hydrogen bombs could thus be carried by rockets already in development or in design. By the spring of 1954 the United States was working toward a hydrogen bomb small enough and light enough to fit in a warhead. In June, heeding the earlier recommendation of a high-level advisory committee, President Dwight Eisenhower

gave top priority to the Air Force's proposed ICBM, named Atlas. There was reason for urgency: intelligence reports revealed that the Soviets were at work on their own ICBM.

Anticipating the ICBM decision, Bollay's group was already working to perfect a more powerful and efficient engine for Navaho that was later adapted for Atlas. One key was switching propellants from alcohol to higher-energy kerosene, the main ingredient of jet fuel. But it was soon clear that jet fuel wasn't good enough for rocket engines; impurities caused a host of

MIG-15 UTI

The primary fighter used by Communist forces in the Korean War, the Russian-built MiG-15 stunned U.S. pilots with its speed and agility. The power plant for this swept-wing fighter was based on an engine made by Britain's Rolls-Royce and shared with the Soviets soon after World War II. Pictured is a two-seater trainer version.

ROCKETS FOR THE COLD WAR

(LEFT) F-86s IN KOREA

The first swept-wing U.S. fighter, the F-86 had as one advantage its tail surfaces, which could be used for control during transonic flight. The other came from its superbly trained pilots: although the swift MiG-15 could outclimb and outrun the American fighter, F-86 pilots in the Korean War scored ten times as many kills as their opponents.

(BELOW) F-86 AT THE MUSEUM

The Museum's F-86 was flown in Korea in combat.

(ABOVE) **RUSSIA'S SPACE WORKHORSE**

Little-changed in four decades, Russia's Soyuz launcher is a direct descendant of the Soviet R-7, the world's first successful ICBM. All thirty-two nozzles fire at liftoff in the Sputnik launcher; they produced a combined thrust of about 900,000 pounds.

(LEFT) **SPUTNIK 1 LAUNCH**

Perched in the nose of an R-7 ballistic missile, the world's first artificial satellite—Sputnik—rises toward space on October 4, 1957.

(ABOVE) **SPUTNIK REPLICA AT THE MUSEUM**

Only twenty-three inches in diameter and weighing 184 pounds, Sputnik contained two radio transmitters to signal its existence. The replica of Sputnik hanging in the Milestones gallery is on loan from the Science in Russia Exhibition of National Achievement.

(LEFT) **SPUTNIK'S PATH**

Circling Earth at 18,000 miles per hour, Sputnik 1's motion against the stars is captured in this time-exposure taken from Plymouth, New Hampshire, two days after its launch.

199

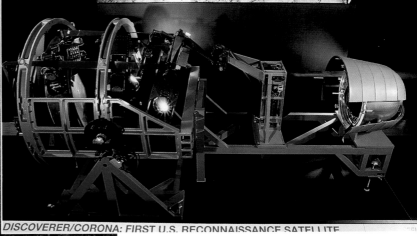

DISCOVERER/CORONA: FIRST U.S. RECONNAISSANCE SATELLITE

(ABOVE) **CORONA CAMERA**

In 1995 the Central Intelligence Agency and the National Reconnaissance Office donated a Corona camera to the Museum. The film spool and camera lenses are on the left; the recovery vehicle for returning film from space to Earth is on the right.

(BELOW) **CORONA RECOVERY (DISCOVERER 17)**

To recover exposed film of targets in the Soviet Union, the Air Force dispatched a Fairchild C-119 or Lockheed C-130 (pictured) to snare the capsule as it descended on its parachute. In its early years the project was known as Discoverer to the public and was billed as a scientific satellite.

(ABOVE) **SURVEILLANCE FROM SPACE**

The first U.S. satellite effort was a secret one. Corona was a photographic reconnaissance satellite first successfully launched in 1960. An August 1960 image of a Soviet airfield, returned from Corona, was among the first satellite reconnaissance photos.

(OPPOSITE) **ATLAS LAUNCH SUCCESS**

In one of its first successful launches, a fully powered Atlas rises on a column of flame in August 1958.

(ABOVE) **X-15 AT THE MUSEUM**

The X-15, designed to probe the limits of rocket-powered atmospheric flight, flew higher and faster than any other airplane. Flying for the Air Force and NASA, the X-15 attained speeds more than six times the speed of sound and soared
as high as sixty-seven miles, far enough to be considered space flight. Three X-15s were built; they made a total of 199
flights. X-15 number 1 hangs in the Milestones gallery.

(OPPOSITE) **B-52 WITH THE X-15 UNDER ITS WING**

The B-52 bomber, workhorse of the U.S. Air Force for more than four decades, could haul more than 75,000 pounds of
bombs over a range of 6,000 miles. The B-52 made its debut in 1955 as a carrier for thermonuclear weapons before the
advent of the ICBM. It also served as the drop aircraft for the rocket-powered X-15 research plane, shown here with an
F-100 chase plane alongside.

CHAPTER SEVEN

RACE TO THE MOON

(ABOVE) **FIRST INTO SPACE**

Yuri Gagarin was a twenty-seven-year-old Russian fighter pilot with relatively little experience when he became the world's first space traveler, piloting the Vostok spacecraft on a single orbit of Earth.

(OPPOSITE) **A WALK IN SPACE**

Floating in the void a hundred miles above Earth, Gemini IV astronaut Ed White takes the first U.S. walk in space, a crucial milestone on NASA's road to the Moon. During his twenty-three minutes outside Gemini IV on June 3, 1965, White tested a hand-held maneuvering gun powered by compressed nitrogen, visible in his right hand. A twenty-seven-foot umbilical, wrapped in gold tape, links White to the spacecraft's oxygen supply and radio. Gemini IV commander Jim McDivitt took this photograph.

April 1961 was not a good month for John Kennedy. The Bay of Pigs fiasco on the nineteenth, when CIA-backed guerrillas failed in an attempt to overthrow the Communist regime of Cuba's Fidel Castro, was bad enough for the new president. But another significant blow had come a week earlier: on April 12 a twenty-seven-year-old Russian pilot named Yuri Gagarin became the first man in space. In a craft called Vostok ("east"), Gagarin made a single orbit of Earth. Only 108 minutes after his launch atop a converted R-7 missile, Gagarin ejected from the descending Vostok and parachuted into southeastern Russia. On his return to Moscow, Gagarin was greeted by a jubilant Nikita Khrushchev, who hailed his flight as a triumph of Soviet Communism. No one in the West knew that Gagarin had been in serious danger from a malfunction just before reentry; his flight was portrayed as flawless. Less than four years after Sputnik, the Soviets had dealt another blow to U.S. morale.

At last, on May 5, Kennedy saw the nation's spirits soar when Mercury astronaut Alan Shepard became America's first man in space by riding a Redstone rocket on a fifteen-minute suborbital hop. Even as Shepard won his own hero's welcome, Kennedy was already planning to launch a colossal effort that would give the United States preeminence in space. If Dwight Eisenhower had been lukewarm about space exploration, Kennedy now embraced it as a contest for the hearts and minds of people around the world. On May 25, standing before a joint session of Congress, he declared, "I believe that this nation should commit itself to achieving the goal, before this decade is out, of landing a man on the Moon and returning him safely to Earth."

If Kennedy was nervous about making such an audacious challenge—and his closest aides could tell that he was—then his words struck some of NASA's space planners as wildly optimistic. Mercury astronauts had yet to orbit Earth, and Kennedy was already calling for a manned lunar landing within eight years. "I thought he'd lost his mind," Chris Kraft, then director of NASA's Flight Operations, said in 1988. "I didn't even think about [the technical requirements]. I just knew it was a job of fantastic magnitude."

Once they had a chance to recover, Kraft and his colleagues realized that with an all-out effort, Kennedy's challenge could probably be met. As John Glenn and his fellow Mercury astronauts reached orbit

211

would have to do it." But the astronauts, like everyone at NASA, were no strangers to the test-flight business, in which caution must always be balanced by acceptance of necessary risk.

The Soviets went even further. Sergei Korolev and his team were under pressure to curtail space activities to free precious funds and manpower for defense programs. To keep the space program alive, Korolev sought to boost its propaganda value with one space spectacular after another. In the process, he took some extraordinary chances. In October 1964 the Soviets bested the Americans' planned two-man flights by orbiting three cosmonauts in a craft called Voskhod ("sunrise"). They returned home after only a day (during which time Khrushchev had been overthrown), but that was enough to stun U.S. observers, some of whom wondered whether the Soviets had matched Apollo's planned capabilities years before Apollo was even ready to fly. The Soviets did not reveal that Voskhod was nothing more than a converted Vostok; to make room for three people in a ship designed for one, the cosmonauts flew without the benefit of space suits or even ejection seats. To some in the West, their flight, coming months before Gemini's manned debut, made the Americans look like also-rans. Nevertheless, when Gemini 3 astronauts Gus Grissom and John Young finally reached orbit on March 23, 1965, NASA could celebrate

a space first of its own: in their five and a half hours aloft, Grissom and Young used their Gemini's thrusters to change the shape and orientation of their orbit.

But Gemini 3's success came only days after yet another Soviet spectacular. On March 18 Pavel Belyayev and Aleksei Leonov reached orbit in Voskhod 2, this time equipped with space suits and a collapsible airlock. Early in the flight Leonov emerged from Voskhod 2 and spent twelve minutes floating in the void. The world's first spacewalk almost ended in disaster, as Leonov struggled in his rigid, bloated space suit to reenter the narrow airlock.

THE FIRST SPACE WALK

Encased in a pressurized space suit, cosmonaut Aleksei Leonov pushes away from Voskhod 2 in history's first spacewalk on March 18, 1965. Mirrored goggles within Leonov's helmet shielded his eyes from the unfiltered solar glare. For almost half of his twenty minutes in the vacuum of space, Leonov struggled to squeeze back into Voskhod 2's narrow airlock.

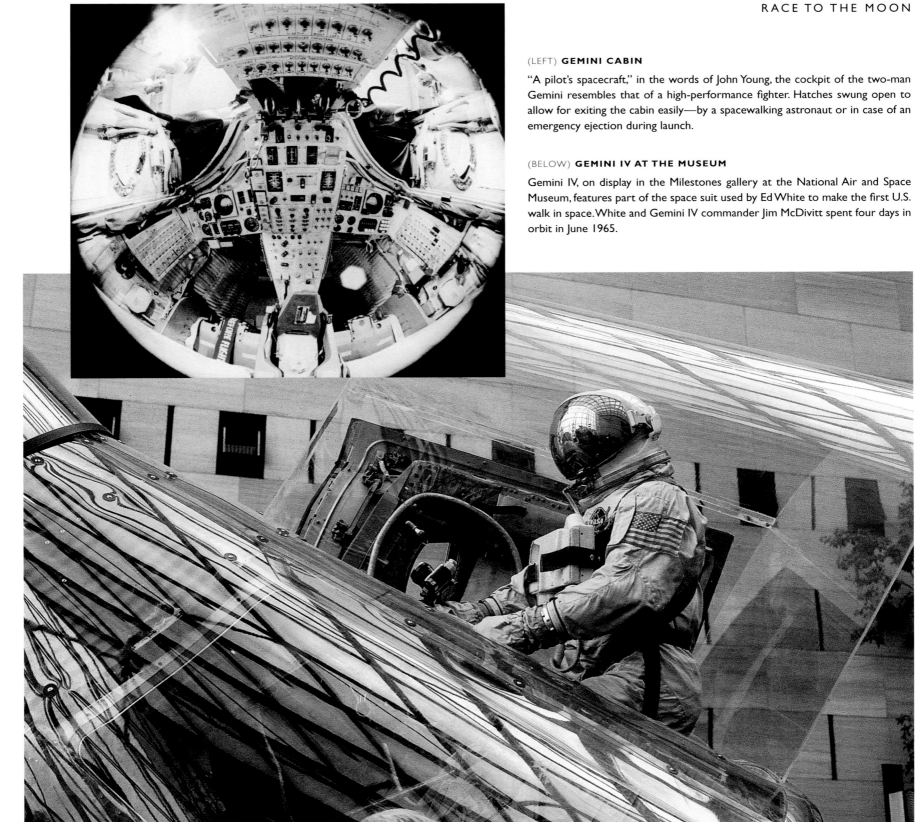

(LEFT) GEMINI CABIN

"A pilot's spacecraft," in the words of John Young, the cockpit of the two-man Gemini resembles that of a high-performance fighter. Hatches swung open to allow for exiting the cabin easily—by a spacewalking astronaut or in case of an emergency ejection during launch.

(BELOW) GEMINI IV AT THE MUSEUM

Gemini IV, on display in the Milestones gallery at the National Air and Space Museum, features part of the space suit used by Ed White to make the first U.S. walk in space. White and Gemini IV commander Jim McDivitt spent four days in orbit in June 1965.

Scott had no choice but to return to Earth at once. They splashed down safely in the Pacific—disappointed, but alive.

NASA solved the thruster problem for future missions, but Gemini's difficulties weren't over. On Gemini IX, spacewalker Gene Cernan tried in vain to test a rocket-powered backpack. Working in weightlessness with no means of adequately restraining his body was like being on a three-dimensional skating rink; each movement sent Cernan tumbling away from his work. Exhausted and overheated, he was forced to abandon his effort and return to the cabin. As Cernan caught his breath, NASA realized that Ed White's trouble-free walk had created a false sense of security. And even as astronauts reached new heights on Gemini X and XI—thanks to the Agena's rocket engine, which boosted them to record altitudes of up to 850 miles—spacewalking remained Gemini's biggest unmet objective, until the final Gemini flight. Thanks to meticulous underwater training and a variety of handholds, tethers, and foot restraints, Gemini XII's Buzz Aldrin spent a total of more than five hours outside with no difficulties in November 1966. Gemini XII's splashdown ended a program that had given the United States a clear lead in space. At NASA all seemed ready for the Moon.

But in 1967 the Moon suddenly seemed to slip from reach. On January 27 the crew of the first manned Apollo flight—Mercury

(OPPOSITE) **GEMINI XI LAUNCH**

The moment that made NASA nervous: launch of the Gemini-Titan II. Ejection seats offered the only means of escape in the event of a malfunction of the 90-foot booster, which delivered 430,000 pounds of thrust at liftoff. Here, Gemini XI astronauts Pete Conrad and Dick Gordon head for orbit on September 12, 1966.

(RIGHT) **GEMINI XII: SPACEWALK SUCCESS**

During one of three spacewalks he made during the Gemini XII mission, Buzz Aldrin photographs himself while standing in the craft's open hatchway. Aldrin's success in more than five hours of space walks removed the last hurdle before Apollo flights could begin.

and Gemini veteran Gus Grissom, space-walker Ed White, and rookie Roger Chaffee—perished during a ground test when a flash fire swept through their sealed command module. Out of the ashes of Apollo 1 came the horrifying realization that because of a number of design flaws, the Apollo 1 command module had been a death trap. No one at NASA or its contractors had intentionally compromised the as-

tronauts' safety. But to some observers, it seemed clear that the pressure of meeting the end-of-the-decade deadline had caught up with them. Looking back, many Apollo veterans would later say that the redesign effort following the fire made the difference between Apollo's ultimate success and failure.

Nor were the Soviets immune to disaster. To mark the fiftieth anniversary of the

(ABOVE) **TRAGEDY STRIKES APOLLO 1**

The blackened interior of Apollo 1 reveals the devastation wrought by the fire. The causes of the tragedy—too many flammable materials in the cabin, pressurization with pure oxygen at sixteen pounds per square inch, and lack of a quick-opening hatch—were remedied when the command module was redesigned.

(RIGHT) **THE CREW OF APOLLO 1**

The crew of Apollo 1 (from left: Gus Grissom, Ed White, and Roger Chaffee) pose before Pad 34 at the Kennedy Space Center, where they were to be launched on a fourteen-day, Earth-orbit test flight. Less than seven weeks after this picture was taken, the three astronauts perished when a flash fire swept through the cabin of their command module during a ground test.

(OPPOSITE) **THE MOON ROCKET: SATURN V LAUNCH**

Rivaling the dawn with its own fire, the first Saturn V moon rocket builds up to 7.5 million pounds of thrust seconds before liftoff on November 9, 1967. The Saturn propelled the Apollo 4 spacecraft to an 8-hour, 37-minute, 8-second test flight that paved the way for manned lunar missions.

Russian Revolution, Soviet leaders were pushing for a manned flight around the Moon by the end of 1967. But that was a long shot at best. For the new Soyuz ("union") spacecraft—an Earth-orbit variant of the Soviet manned lunar ferry—engineers had expended great effort toward automation, reducing the cosmonauts' piloting role. That move proved costly, in both development time and complexity, and several unmanned Soyuz tests were plagued by failures. In April 1967,

over some planners' reservations, the decision was made to launch Vladimir Komarov into orbit aboard Soyuz 1. Problems with a power-producing solar panel forced controllers to order Komarov back to Earth. A planned linkup with a second Soyuz had already been canceled. After Komarov guided Soyuz 1 through a difficult manual reentry when the automatic system malfunctioned, Soyuz 1's parachute became hopelessly tangled and the craft slammed into the ground, killing

Komarov instantly. The Soviets could now reflect on what Gus Grissom had said only weeks before the Apollo fire: "If we die, we want people to accept it. We're in a risky business."

For NASA, at least, 1967 ended well. On November 9 the giant Saturn V Moon rocket—the crowning creation of Wernher von Braun and his team—towered 363 feet above its Cape Kennedy launch platform like a monument to human audacity. The sheer enormity of the effort required to create it was as staggering as the rocket itself. Hundreds of separate contractors around the country had built and tested its 3 million parts and components. With a total chemical energy equivalent to a small atomic weapon—7.5 million pounds of thrust at liftoff, a hundred times more than that of the Redstone that launched Alan Shepard—the Saturn V was the most powerful flying machine ever perfected. When it ascended into a cloudless sky, its deafening shock wave shook buildings and pounded at the awed spectators. Eight hours later, its payload—the unmanned Apollo 4 command module—splashed down to end NASA's year on a note of success.

In the spring of 1968, however, there was bad news. The nation was torn by the assassinations of Martin Luther King Jr. and Robert Kennedy, dissent on college campuses over the war in Vietnam, and unrest on city streets. And there were more

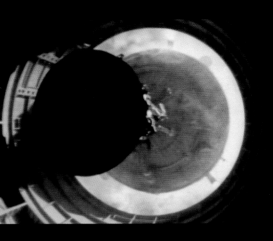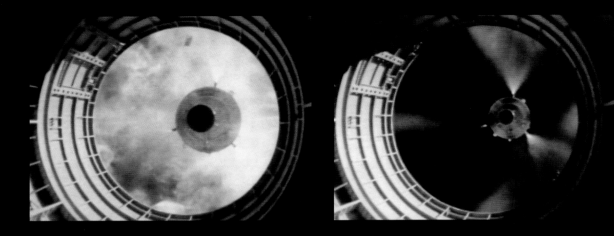

(ABOVE) HEADING FOR ORBIT

The second stage of a Saturn IB booster ignites and propels the Apollo 5 spacecraft toward Earth orbit during an unmanned test flight on January 22, 1968. The scene was filmed by an automatic camera inside the Saturn's first stage.

(RIGHT) LIVE FROM APOLLO 7

October 14, 1968: Thanks to an onboard television camera, earthbound viewers were treated to a glimpse of Apollo 7 astronauts during their flawless, eleven-day flight. At left is Donn Eisele; mission commander Wally Schirra (right) indulges his penchant for levity with a sign reading, "Keep those cards and letters coming in, folks."

For NASA managers, intelligence reports of the impending flight had added impetus to their own lunar-orbit decision. But the Soviets had not forgotten the trauma of losing Vladimir Komarov. Now, after several malfunctions on unmanned circumlunar test missions, they refused to give the go-ahead for Leonov and Makarov's flight. The window of opportunity for the launch came and went, and the two disappointed cosmonauts waited to see if the Americans would make the journey.

In the early morning of December 21, at Florida's Kennedy Space Center, Frank Borman and his crew prepared to leave Earth. The first men to ride the Saturn V were awed by its power—the rocket shook them in their harnesses as it climbed away from its launch platform—and thankful for its perfection. Eleven minutes after liftoff the men were in orbit. Less than three hours later the Saturn's third-stage engine re-ignited to propel Apollo 8 onto a course for the Moon. At that point, all similarities with previous space flights ended. Through the command module's small windows, the men could see Earth as a radiant globe dwindling in the blackness. Conversations with mission control were punctuated by noticeable silences while their radio signals traversed the increasingly vast distance. The moonward voyage was marred only by Borman's bout of motion sickness that doctors would later

blame on weightlessness's effects on the inner ear.

Sixty-nine hours after starting out, Apollo 8 arrived at its goal and slipped behind the 2,160-mile-wide sphere, out of contact with Earth. Over the lunar farside, a four-minute blast of the CSM's main rocket engine put Apollo 8 in orbit. Then all was still, and the men looked down at a bleached and utterly desolate landscape pockmarked by craters of all sizes. For twenty hours they circled as low as sixty-nine miles. Borman monitored the spacecraft, Anders took photographs and studied the landscape, and Lovell sighted on lunar landmarks using an onboard sextant. Apollo 8 would return with a wealth of data for the lunar landing; it would also give scientists a taste of the explorations to come. But the most precious cargo would turn out to be a single photograph of Earth, an oasis of life and color, rising beyond the Moon's battered, barren face.

On that Earth, where it was Christmas Eve, untold millions shared in the

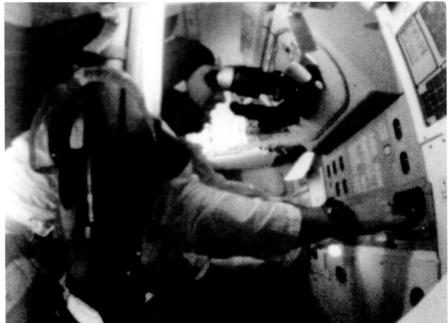

SPACE NAVIGATOR

On the way back to Earth, Jim Lovell sights through Apollo 8's onboard navigation optics. If they had lost radio contact with mission control, the astronauts would have used their own navigation sightings to stay on the right path for reentry.

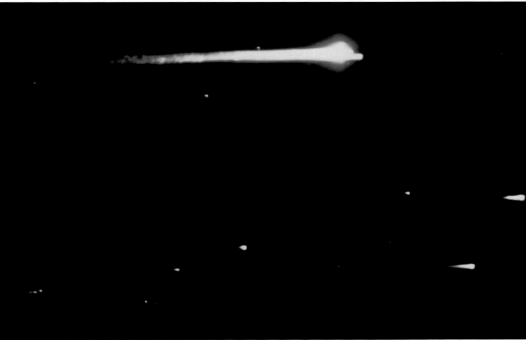

exploration, thanks to live black-and-white television transmissions from Apollo 8. And they heard a message that crystallized the impact of this first lunar voyage, as the astronauts read the first ten verses of Genesis. Then it was time for Borman's crew to leave. In Houston, flight controllers sweated out Apollo 8's reemergence from behind the Moon, knowing that the CSM's engine had to work or the astronauts would be stranded in lunar orbit with no hope of rescue. At last they heard Jim Lovell's voice: "Please be informed there is a Santa Claus." On December 27, after a precise high-speed reentry, the first Moon voyagers splashed down safely in the Pacific. As 1969 opened, a surging confidence took hold of the Manned Spacecraft Center; suddenly the lunar landing was within reach.

If Apollo 8's spectacular success dealt a body blow to the Soviet lunar effort, then the failure of their giant N-1 Moon rocket all but ended it. The N-1's first test launch in February ended seconds after liftoff when the mammoth rocket exploded. (A second test in July would also end in disaster.) Publicly, the Soviets now denied there had ever been a race at all and stressed the importance of their Soyuz missions in Earth orbit. Meanwhile, NASA scored two more victories with the flights of Apollo 9 and 10, full-up tests of the CSM and LM in Earth orbit and lunar orbit, respectively. Incredibly, these two missions, each far more complex and demanding than Apollo 8 had been, accomplished all their objectives. The way was now clear for Apollo 11 to attempt Apollo's climactic mission.

Within a year after John Kennedy launched the United States on a course for the Moon in 1961, Soviet planners began thinking about how to get there first. As in the United States, Russian scientists had already worked out much of the theory behind a Moon flight. Engineers had even sketched proposals for lunar missions. Not until 1964, however, did the Kremlin grant official approval to the Moon program; the Soviets were behind even before they could begin. In addition, resources were limited, in both funding and manpower; the same design bureaus charged with building the lunar spacecraft were supplying missiles for the military.

By the mid-1960s, two separate lunar programs were under way, one to send cosmonauts around the Moon without orbiting, and one for the lunar landing. Both used hardware based on the Soyuz that was then being readied for its first flights. A modified Soyuz called Zond, equipped with an upgraded communications system and heat shield, was to be used for the circumlunar flight. For the landing, as with the U.S. Apollo, there were two spacecraft. A ferry craft called the *Lunniy Orbital'ny Korabl'* (lunar orbital ship, or LOK), was also based on the Soyuz design. Equipped with a special rocket, the LOK would carry two cosmonauts to and from lunar orbit.

The Moon lander, called the *Lunniy Korabl'* (lunar ship, or LK), bore some resemblance to the Apollo lunar module but was roughly 25 percent shorter and only half as wide. Like its American counterpart, the LK had two sections, the bottom one equipped with landing legs and a ladder.

ILL-FATED N-1 LAUNCH

Its thirty first-stage engines delivering 9.9 million pounds of thrust, the N-1 booster rises from its launch pad at the Baikonour Cosmodrome in central Asia. Seconds later, the rocket began to roll violently, then disintegrated. This was the third of four failed test launches of the giant rocket, which would have been the booster for the Soviet manned lunar landing missions.

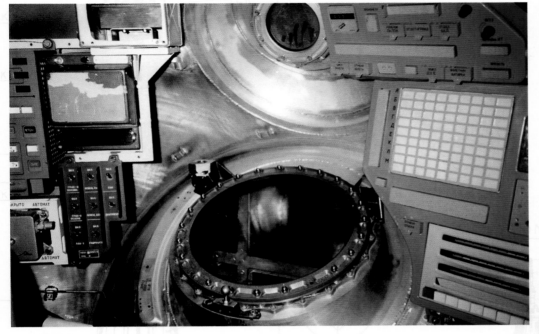

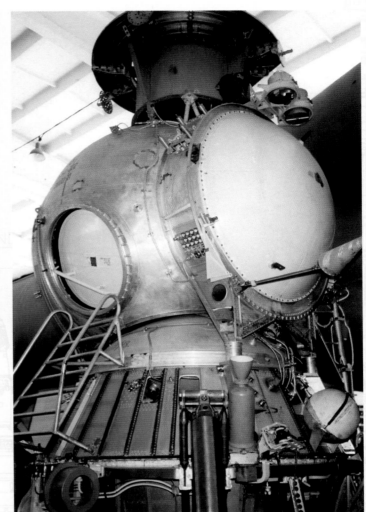

(ABOVE) **CABIN OF THE LK**

Inside the LK, controls and displays were to be used by a lone cosmonaut in landing on the Moon.

(RIGHT) **THE SOVIET MOONSHIP**

The Soviet Moon lander, officially called *Lunniy Korabl'* (lunar ship), rests in the Moscow Aviation Institute. Meant to transport a single cosmonaut from lunar orbit to the surface and back, the LK was to be flown almost entirely under automatic control. The same rocket used for final descent was also to be used for ascent.

(IN BACKGROUND) **N-1**

Drawings of the N-1 by Charles Vick.

However, it was designed to carry only one cosmonaut to the surface of the Moon.

The top section contained controls, life support, and communications for a lone cosmonaut, as well as a rocket engine for the return trip from the Moon's surface to lunar orbit.

Many details of the lunar landing mission plan are still uncertain, but a general picture has been pieced together from accounts by Soviet space engineers, studies by Western researchers, and a variety of other sources. A giant N-1 booster would be used to propel the LOK, the LK, and a special two-stage lunar rocket system into Earth orbit. One stage of the lunar rocket system would propel the cosmonauts onto a course for the Moon, then be cast off. The men would use the second stage to refine their path and then, after arriving at the Moon, fire it again to go into lunar orbit.

Circling the Moon, the mission commander would don his lunar space suit and leave his companion in the LOK, making a space walk to the LK. The LOK would then separate and remain in orbit, while the LK and its attached rocket stage descended toward the Moon. The rocket stage would brake the LK's descent until within an altitude of one to three kilometers; then it would be jettisoned, and the LK's own rocket engine would take over. Finally, the commander would steer the LK to a touchdown on the Moon.

With this achievement behind him, the mission commander would find that his work was just beginning. After making necessary preparations, he would emerge from the LK, climb down the ladder, and spend several hours exploring the Moon's surface. Back inside, he would fire the LK's engine, leaving the craft's bottom section on the Moon, and head for a rendezvous with his companion in lunar orbit. A docking probe in the nose of the LOK would attach to a perforated plate on the LK's roof, joining the two craft. Then the commander would make yet another space walk back to the LOK.

The rest of the mission would proceed much like an Apollo lunar flight: the two cosmonauts would cast off the LK's upper section, then fire the engine in their LOK to leave lunar orbit and begin the journey home. Days later the LOK's crew module would become the only portion of the spacecraft to return to Earth's surface, slowed by parachutes and retrorockets to a landing.

No one can say what would have happened if failures of the N-1 booster had not ruled out the Soviets' Moon landing plans. In retrospect, however, the challenges of the planned mission seem extremely daunting. This is especially true for the commander, who would have had to carry out the lunar landing, two spacewalks, one moonwalk, and a lunar liftoff and rendezvous, without assistance. Interestingly, Soviet planners were far more concerned about the reliability of the hardware; they gave the LK something the Apollo lunar module didn't have: a backup rocket engine for getting off the Moon's surface in case the primary engine failed.

SUIT FOR A SOVIET MOONWALKER
The Krechet space suit, developed for a moonwalking cosmonaut, featured a hard torso with a built-in backpack that was hinged to allow entry into the suit.

SOYUZ IN EARTH ORBIT

The Soyuz spacecraft made its first successful flight in 1968. For their planned Moon missions, cosmonauts intended to fly a modified version of Soyuz to and from lunar orbit. Shown here is a Soyuz variant flown for the joint U.S.-Soviet mission of 1975. A more recent variant—the Soyuz-TM—remains in use today.

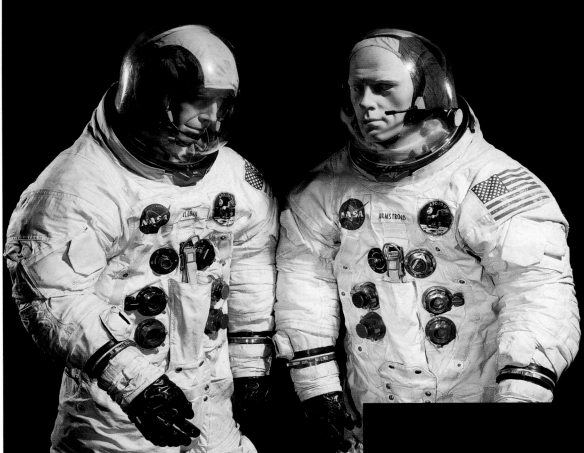

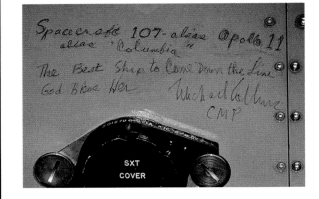

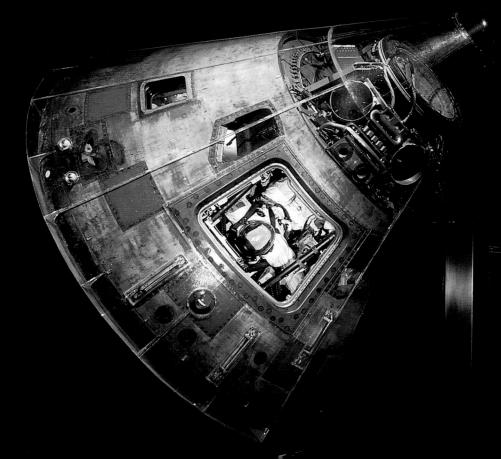

(ABOVE) **ARMSTRONG'S AND ALDRIN'S SPACE SUITS**

The space suits worn by Neil Armstrong and Buzz Aldrin on Apollo 11 are on display in the Apollo to the Moon gallery. They are seen here in the configuration used during Apollo 11's launch. Blue and red connectors on the chest are for oxygen hoses, communications, and cooling water. The astronauts left their lunar backpacks and boots on the Moon's surface.

(ABOVE, RIGHT) **"THE BEST SHIP"**

Soon after splashdown, Collins inscribed on a wall of the spacecraft's cabin, "Spacecraft 107—alias Apollo 11—alias *Columbia*. The best ship to come down the line. God Bless Her. Michael Collins, CMP [command module pilot]."

(RIGHT) *COLUMBIA* **AT THE MUSEUM**

The Apollo 11 command module *Columbia* in the Milestones gallery. The space suit worn by Apollo 11 command module pilot and former Museum director Mike Collins is visible through the side hatchway.

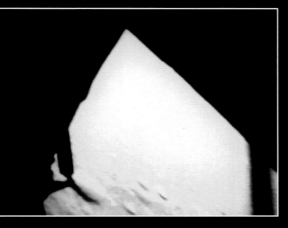

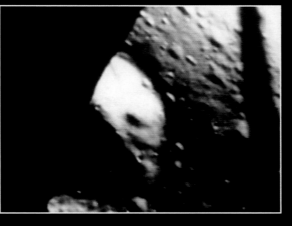

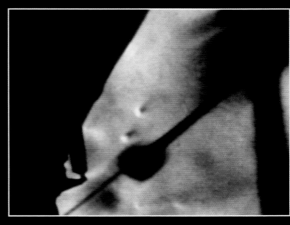

(ABOVE) FINAL MINUTES OF THE APOLLO 11 LANDING

(LEFT) A movie camera on board *Eagle* recorded these views during the descent to the Moon. At 20,000 feet, a bright horizon is visible.

(MIDDLE) At 180 feet, *Eagle* flies over an 80-foot-diameter crater, heading for safer ground.

(RIGHT) With just 10 feet to go, the shadow of one of *Eagle*'s landing legs comes into view. Dust kicked up by the lunar module's descent rocket veils the surface.

(RIGHT) A MESSAGE FROM EARTH

Photographed during the Apollo 11 moonwalk, the aluminum plaque attached to the front leg of the lunar module *Eagle*—left on the Moon—reads, "Here men from the planet Earth first set foot upon the Moon, July 1969 A.D. We came in peace for all mankind."

(LEFT) THE VIEW TOWARD HOME

From the Arabian Peninsula to the Antarctic icecap, the full Earth is photographed by the Apollo 17 astronauts as they head moonward during the final lunar landing mission on December 7, 1972.

(OPPOSITE, TOP) FAREWELL TO THE MOON

Ending the first brief age of human lunar exploration, the Apollo 17 Lunar Module *Challenger* leaves the Moon on December 14, 1972. A television camera on the Lunar Rover, remotely controlled from Earth, captured these views of *Challenger*'s liftoff.

(OPPOSITE, BOTTOM LEFT) THE LAST MAN ON THE MOON

The last man to leave his tracks in lunar dust, Apollo 17 commander Gene Cernan stands in the Moon's Taurus-Littrow valley on December 13, 1972.

(OPPOSITE, BOTTOM RIGHT) CERNAN'S SPACE SUIT IN THE APOLLO TO THE MOON GALLERY

Cernan's space suit, on display in the Apollo to the Moon gallery, is grimy after three days of exploring the lunar surface. Moon dust, made of tiny particles of rock and natural glass, worked its way into the suit's beta-cloth outer layer.

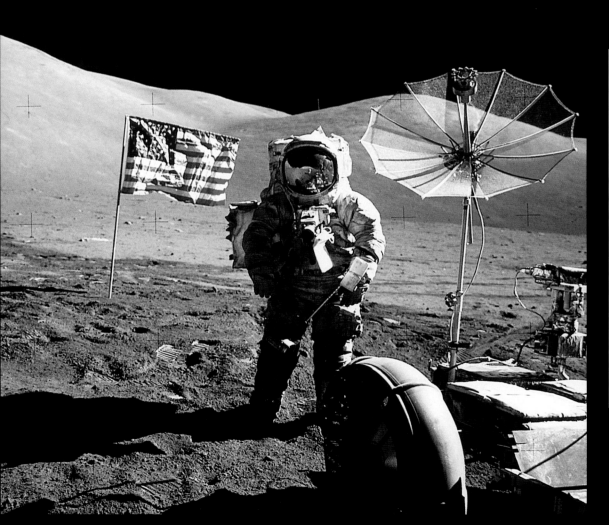

(ABOVE) SKYLAB SPACE STATION

The Skylab space station, launched in 1973, offered its three-man crews as much living and working space as a small house. The longest mission to the Orbital Workshop, lasting eighty-four days, ended in 1974. Five years later Skylab reentered Earth's atmosphere; pieces of the craft were found in Australia.

(AFTER GATEFOLD, PAGE 259) SKYLAB AT THE MUSEUM

The backup Skylab, capable of its own missions but never launched, rests in the Museum's Space Hall.

SPACE PROGRAMS OF OTHER NATIONS

(LEFT) EUROPE'S ARIANE BOOSTER

The European Space Agency's series of Ariane boosters, first launched on December 24, 1979, have made Europe a leader in commercial space activities. Shown is a launch of the Ariane LO-3 in 1981.

(BELOW) JAPANESE SATELLITE LAUNCH

Japan's space program, in addition to commercial satellites, has emphasized scientific exploration. Shown here is the launch of Japan's first interplanetary spacecraft. *Sakigake* ("Pioneer") was launched January 8, 1985, on a MU-3S2 rocket.

(RIGHT) CHINA'S LONG MARCH 2E

One of the mainstays of the Chinese space program is its Long March series of satellite launchers. Long March 1 carried the China-1 satellite into orbit on April 24, 1970. Here, China's Long March 2E is displayed at a 1996 air show.

Shuttle would actually turn a profit. Such promises—along with election-year concerns of the White House and Congress about aerospace jobs in California and Texas—secured the Shuttle's approval.

The House vote came in April 1972 as Apollo 16 astronauts John Young and Charlie Duke explored the Moon's Descartes highlands. Eight months later the splashdown of Apollo 17 closed NASA's brief era of manned lunar exploration. If the end seemed premature to many in the Apollo teams, many others in NASA were eager to move on. The Skylab marathons were slated to begin in 1973. A new détente between the United States and the Soviet Union had spawned plans for a joint space mission in 1975. And if all went according to plan, the first Shuttle would reach orbit in 1978. During the Shuttle's development there would be a lengthy hiatus in manned flights, but NASA hoped it would be worth the wait.

The Soviet space program underwent its own redirection. For a time, a die-hard cadre of engineers fought to keep the lunar-landing effort alive; they talked of two-week visits and a lunar base. More failures of the N-1 booster ended their hopes; the Soviets officially abandoned the Moon in 1974. But cosmonauts were already expanding their presence in Earth orbit. In June 1971, two years ahead of NASA's first Skylab mission, the three-man crew of Soyuz 11 had logged three weeks

259

Mars orbit for a thorough reconnaissance and revealed a world of astounding geologic variety, with towering, extinct volcanoes, a vast network of canyons, and winding channels bearing an intriguing resemblance to dry river valleys. Mariner's Mars was nothing like the inhabited world envisioned by Percival Lowell at the turn of the century. But it tantalized scientists with the hope that some form of life, however simple, might exist there. In 1975 a pair of Viking orbiters and landers, the

DESTINATION: EARTH-ORBIT

A team of Soviet cosmonauts lifts off in a Soyuz-TM spacecraft, headed for the Mir space station. In the decades after the Moon race, cosmonauts made record-breaking stays aboard orbiting space stations as part of a methodical assault on the limits of space endurance.

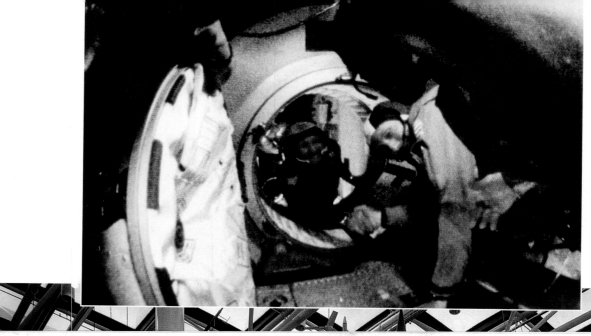

(ABOVE) *747 FERRYING THE SHUTTLE*

The Boeing 747, centerpiece of the 1970s' wide-body revolution in commercial air transport, also serves as a ferry craft for NASA's Space Shuttle orbiter. Here it carries the *Enterprise,* which astronauts piloted on a series of approach and landing tests in 1977. *Enterprise* is now part of the Museum's collection.

(OPPOSITE) **THE SPACE SHUTTLE'S MAIDEN VOYAGE**

April 12, 1981: The Space Shuttle *Columbia* lifts off carrying John Young and Bob Crippen on the Shuttle's maiden voyage. The two solid rocket boosters, together with the orbiter's three liquid-propellant engines, deliver a total thrust of 6.9 million pounds.

external tank, the only throwaway element of the Shuttle system, would fuel *Columbia*'s three main engines during launch. Those engines—designed to be light-weight, to withstand extremely high operating pressures, and, of course, to be reusable—were the source of the Shuttle's knottiest development problems. Two strap-on solid rocket boosters would each deliver an additional 2.6 million pounds of thrust. Five onboard computers would monitor *Columbia*'s systems and fly the spacecraft through its hypersonic reentry. Landing the Shuttle would not be easy; it would descend to Earth as an unpowered glider, with relatively poor lift characteristics, and the astronauts would have only one chance to bring it down safely. The Shuttle's complexity was appropriate for a research-and-development vehicle, but NASA was committed to making it into an operational spaceliner. And to up the ante, the Shuttle could not be tested unmanned. For the first time in NASA's history, astronauts would ride an unproven vehicle into space. As with Gemini, only ejection seats offered a chance for escape in a launch emergency—and only on the first four test flights; they would be removed once the Shuttle was declared operational.

On April 12, 1981—twenty years to the day after Yuri Gagarin's flight—*Columbia* roared flawlessly into orbit carrying veteran John Young and rookie Bob Crippen. Concern mounted when the astronauts discovered that some of the heat-protective

able to meet the breakneck schedule it had promised more than a decade earlier.

The pressure was on to make 1986 even busier and more productive, with Shuttles delivering such important payloads as the Hubble Space Telescope and the Galileo Jupiter probe. There was even a program to let ordinary citizens fly in space. The seven-member crew for the Shuttle *Challenger*'s January 28 launch included a New Hampshire schoolteacher named Christa McAuliffe, who planned to share her experiences with children on Earth in televised lessons. If the Shuttle had not made spaceflight economical, it had made it routine— or so it seemed until *Challenger*'s flight ended in disaster seventy-three seconds after liftoff. Months of investigation by a presidential commission traced the cause of the tragedy to a seal in one of the solid rocket boosters that failed, partly because of cold temperatures on launch day. More fundamental, the commission found, were failures in judgment by NASA and contractor personnel who approved the launch decision despite warnings from some Shuttle engineers.

According to Boston College sociologist Diane Vaughan, who spent nine years analyzing the causes of the *Challenger* tragedy, these errors did not result from hubris or disregard for safety but from a change in NASA's culture. In the 1970s and 1980s, Vaughan says, the space agency was transformed into a vast bureaucracy in

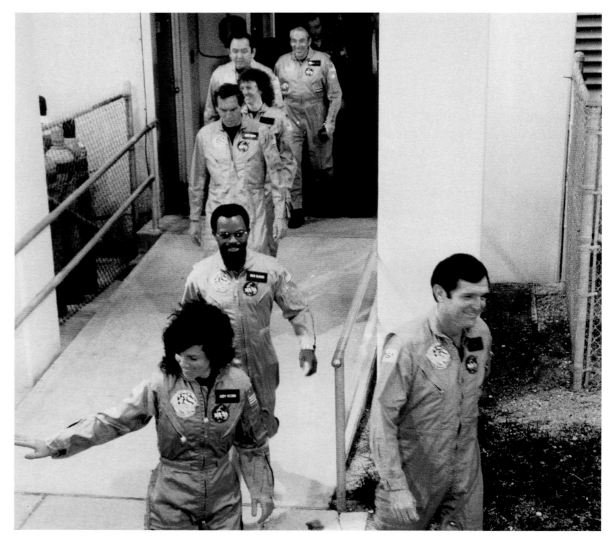

(ABOVE) ***CHALLENGER* CREW HEADING FOR LAUNCH PAD**

The ill-fated crew of the Shuttle *Challenger* heads for the launch pad on January 28, 1986. From right: mission commander Dick Scobee, Judy Resnik, Ron McNair, Mike Smith, Christa McAuliffe, Ellison Onizuka, and Greg Jarvis.

which Shuttle engineers, who were burdened by massive paperwork and schedule pressure, no longer played central roles in the decision-making process. Many of the people who did make decisions failed to perceive the risks they were accepting in letting the Shuttle fly. The Shuttle organization had become so complex that managers did

THE WAR IN VIETNAM

In the mid-1960s and early 1970s, as televised images of the Vietnam War appeared in living rooms across the United States, Americans witnessed the first war where virtually every operation used aviation. Almost everyone who participated in the Vietnam War was affected by aviation in some way. Most Americans who left home to fight a war halfway around the world reached Vietnam by air. Once there, soldiers were shuttled to and from the battlefield by helicopters. Supplies were airlifted on giant cargo planes, on a scale never before possible. And while hundreds of thousands of troops fought the ground war, bombers filled the skies in massive raids: fighters and fighter-bombers flew thousands of sorties from carriers and air bases in Southeast Asia.

During the course of the conflict, the longest in U.S. history, the technology of the air war changed dramatically. At the start, the aircraft in service included World War II–era propeller-driven planes like the Douglas A-1 Skyraider, as well as veterans of the Korean conflict, such as the Douglas A/B-26 Invader. Even a military version of the DC-3, called the AC-47, was equipped with Gatling guns and used for low-level ground attacks. At the same time, workhorses of the 1950s like the B-52 bomber and the C-130 cargo plane became mainstays of the air war. Jet fighters developed just before the start of the war, including the F-100 Super Sabre and the F-4 Phantom, were also used extensively. By the last years of the war, a handful of new-generation aircraft had joined the fighting, including the swing-wing F-111.

Meanwhile, the weapons used in Vietnam included surface-to-air missiles, requiring new advances to protect pilots. Many fighters were outfitted with a new electronic sensors designed to alert pilots to missiles that were still miles away. Sophisticated radar systems helped a bomb-laden plane lock onto its targets. The effect of these innovations was an almost bewildering stream of information in the cockpit. Some pilots switched off their readouts before getting into the thick of combat, preferring to fly the "old-fashioned" way.

(ABOVE) **B-52**

The B-52 Superfortress was used extensively for bombing raids during the Vietnam War.

(RIGHT) **MiG-17**

In a gun-camera view made from an F-105 Thunderchief fighter-bomber, a MiG-17 erupts in flames after being hit by the American's 20mm cannon. The confrontation took place as the F-105 was returning from a bombing raid northeast of Hanoi.

Heavy-lifting helicopters, able to carry increasingly greater loads, thanks to improvements in small gas-turbine engines, were a major innovation and saw greater service in Vietnam than in any previous war. Many were used to carry the wounded from the front lines to Army hospitals. The ability to conduct swift evacuations saved many soldiers' lives.

Ultimately, however, the effectiveness of the array of technologies employed in the air war in Vietnam is still debated among historians.

(ABOVE) DOUGLAS A-4C SKYHAWK

Despite its small size, the Douglas A-4C Skyhawk could carry a prodigious load of armaments, up to 10,000 pounds of bombs. A favorite of U.S. and foreign pilots, this agile attack-bomber saw action in Vietnam, the Middle East, and the Falkland Islands. It holds the distinction of the longest production run for a U.S. military jet, roughly a quarter century.

(LEFT) F-4 PHANTOM

One of the mainstays of the Vietnam era, the F-4 Phantom proved amazingly versatile. Designed as a Navy interceptor, it went on to serve as a strike fighter, long-range fighter, attack bomber, and reconnaissance plane.

FLIGHT RESEARCH CONFIRMS
NEW DESIGN THEORIES

The X-15:

(ABOVE) HAWKER-SIDDELEY FGA MKI KESTREL

One of a series of aircraft taking an innovative approach to vertical flight, the Hawker-Siddeley Kestrel used rotating nozzles to redirect the thrust of its jet engines. The Museum's Kestrel was used by NASA for flight-testing from 1967 to 1974. The first Kestrel flew in 1964.

(LEFT) MiG-21 AT THE MUSEUM

Often likened to the U.S. F-4, the MiG-21 was a versatile and effective fighter used by the Soviet Union and its allies. It is shown here, before restoration, at the Museum's Garber Facility.

F-16s OF THE AIR FORCE THUNDERBIRDS
Computer technology is at the heart of the F-16 Fighting Falcon, seen here in flight with the Air Force's Thunderbirds exhibition team.

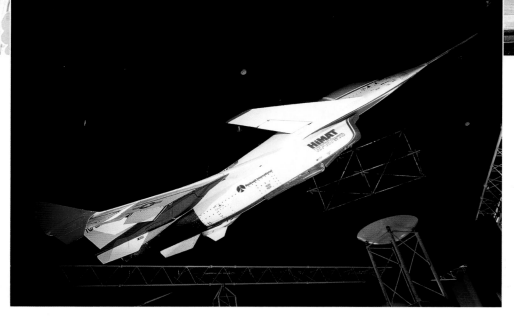

(ABOVE) *VOYAGER* AT THE MUSEUM

Piloting *Voyager*, a composite-built aircraft with a wingspan greater than that of a Boeing 727, Dick Rutan and Jeana Yeager made history's first flight around the world without refueling, in December 1986. Essentially a flying fuel tank, *Voyager* contained just over 7,011 pounds of fuel at takeoff, more than 72 percent of the craft's total weight. It used all but 106 pounds of that fuel during the nine-day flight. In addition to the difficulties of flying the fuel-laden craft, Rutan and Yeager had to contend with violent weather and engine trouble. *Voyager* now hangs inside the Museum's South Lobby.

(LEFT) HiMAT

An unmanned, remote-piloted vehicle developed for NASA, the Rockwell Hi-MAT (Highly Maneuverable Aircraft Technology) featured a variable-camber wing design. The 22-foot long craft, which first flew in 1979, hangs in the Museum's Beyond the Limits gallery.

(LEFT AND BELOW) **THE X-29**

Seen at NASA's Dryden Flight Research Center (left) and as a full-scale mockup in the Museum's Beyond the Limits gallery (below), the Grumman X-29 was designed with the aid of supercomputers and first flew in 1984. Although its radical forward-swept wing layout is highly efficient, it made the X-29 one of the most unstable aircraft flown up to that time; the craft had to be controlled with the help of onboard computers.

CHAPTER NINE

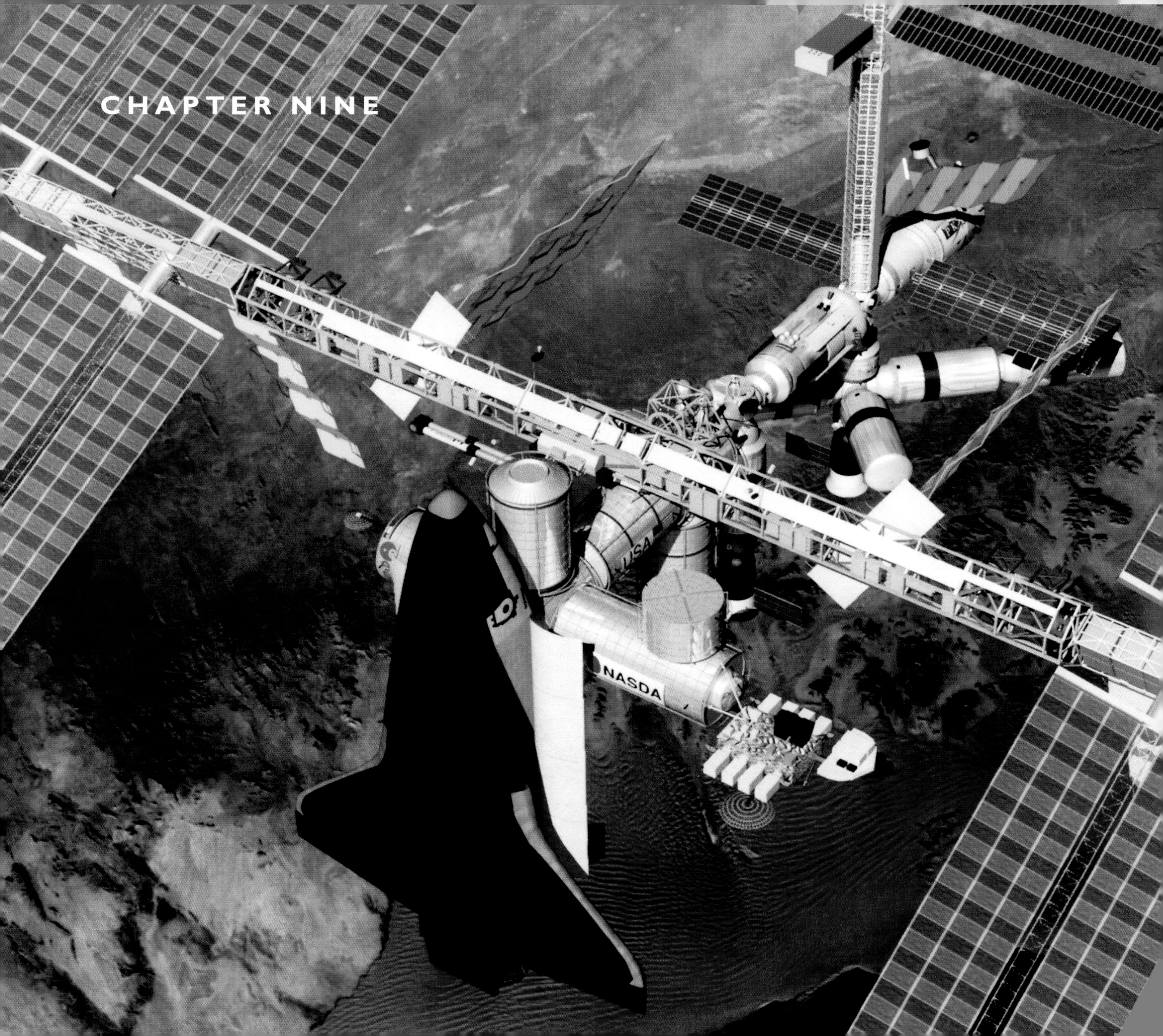

THE ENDLESS SKY

Orville Wright was once asked what the future might hold for aviation. "I cannot answer," he replied, "except to assure you that it will be spectacular." In 1944, at age seventy-two, Orville had a chance to experience some of the truth of his prediction. Lockheed had just created the C-69 Constellation, the military transport version of the four-engine aircraft that would later become a milestone of commercial air transport. On one of the C-69's early flights, Orville was invited to ride in the cockpit. A photograph taken during the flight shows him sitting in the copilot's seat, smiling happily. For a few minutes Orville took the controls; it was the last time he flew before his death in 1948.

(RIGHT) ORVILLE WRIGHT'S LAST FLIGHT

Seated in the cockpit of a Lockheed C-69 Constellation, a smiling Orville Wright is pictured in 1944. During an early flight of the C-69, Orville took the controls for a few minutes: it was the last time he flew before his death in 1948.

(OPPOSITE) AN OUTPOST IN ORBIT

Circling Earth at a height of 250 miles, the International Space Station (ISS) is depicted as it will look in its completed, fully operational state, docked to a Space Shuttle orbiter. The ISS is a joint project of the United States, Europe, Canada, Japan, and Russia and is designed to include seven laboratory modules along with living quarters for a crew of six. Construction in space on the ISS is expected to start in 1998.

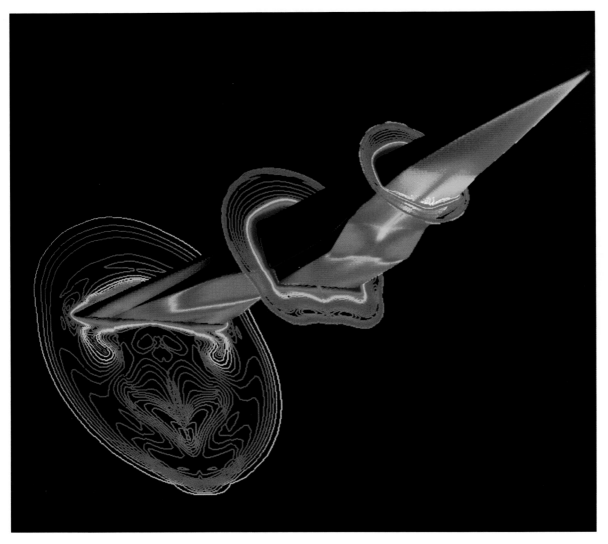

Designed to be completely reusable, the SSTO would attempt to do what NASA could not achieve with the Space Shuttle: slash the cost of access to low Earth orbit. In July 1996 NASA awarded Lockheed Martin a $960 million contract to develop a half-scale, suborbital version of VentureStar that is officially called the X-33. With test flights slated for 1999, NASA hopes the X-33 program will demonstrate the technologies required for a real SSTO. But the Skunk Works' task is monumentally difficult. Only 11 percent of a successful SSTO's total weight can be allotted for the payload, the crew, and the vehicle itself; the remaining 89 percent must be reserved for fuel. To that end, the X-33 prototype will employ a host of new materials, including lightweight composites. In addition, the body of the craft is aerodynamically shaped to generate most of the lift. This "lifting body" approach, which has been tried in small experimental vehicles but never in an operational spacecraft, eliminates the need for large, heavy wings. To power the X-33, engineers have chosen an unusual engine called a linear aerospike. Unlike conventional rocket motors, an aerospike engine doesn't have a bell-shaped exhaust nozzle; instead, it utilizes the flow of air around the speeding rocket to shape the exhaust gases. This approach, say designers, gives the aerospike engine maximum efficiency at any altitude. If all goes well during suborbital test flights in

(ABOVE) **HIGHER AND FASTER**

Engineers have long dreamed of a passenger transport capable of flying at hypersonic speeds and great altitudes, even at the fringe of space. Here, in a 1992 simulation by a NASA supercomputer, a proposed hypersonic vehicle called the National Aerospace Plane flies at ten times the speed of sound—some 6,500 miles per hour.

(OPPOSITE) **VENTURESTAR**

The proposed follow-on for the Space Shuttle, the Lockheed Martin VentureStar, is shown as it would look in Earth orbit. A half-scale, suborbital version has been slated for test flights in 1999. Building the full-scale VentureStar would mean conquering one of the most difficult challenges in space technology: a single-stage (i.e., without booster rockets), reusable orbital vehicle.

MARS SAMPLE-RETURN

Carrying a precious sample of martian rock and soil, a robotic explorer launches itself toward Earth. One of the cherished goals of planetary exploration, the Mars sample-return mission has been scheduled to begin in 2005. Painting by Pat Rawlings.

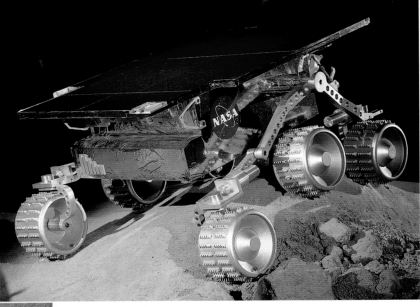

(ABOVE) MARS ROVER

When the Mars *Pathfinder* spacecraft left Earth, heading for a landing on the red planet on July 4, 1997, it carried a 23-pound rover called *Sojourner*. This full-scale model of the rover is on display in the Museum's "Where Next, Columbus?" gallery.

(LEFT) HUMANS ON MARS

Ultimately, human beings may set foot on Mars and establish a base there, as depicted in the "Where Next, Columbus?" gallery. Mars explorers will hunt for evidence of past and present life on the red planet and evaluate its potential as a future home for humanity.

(LEFT) SATURN AND ITS MOONS

Resplendent in this 1981 image from Voyager 2, ringed Saturn and its moons (three of which are visible here) are the target of NASA's Cassini probe, which is scheduled to arrive in the summer of 2004.

(BELOW) LANDING ON TITAN

December 2004: The European Space Agency's Huygens probe, part of the Cassini mission, will parachute into the atmosphere of Saturn's largest moon, Titan. Shrouded in an opaque haze rich in complex organic molecules, Titan may provide clues to the processes that gave rise to life on Earth.

(LEFT) THE ALBERT EINSTEIN PLANETARIUM

Visitors to the Museum's Albert Einstein Planetarium experience the wonders of the night sky with the aid of this Zeiss VIa planetarium projector. The projector, which can display more than 9,000 stars and other celestial objects, was a gift from the Federal Republic of Germany in 1976 for the American Bicentennial and was donated to the Museum by Congress.

(RIGHT) THE SUN'S NEIGHBORHOOD

This stellarium, a scale model of the sun and its stellar neighbors, graces the Museum's "Where Next, Columbus?" gallery. The stars shown lie within 50 light-years of Earth. This is the volume of space reached by terrestrial radio broadcasts during the last half century—but it is only 1/2000 the diameter of our Milky Way galaxy.

(ABOVE) **THE STARSHIP *ENTERPRISE***

Boldly going where no one has gone before, the fictitious space travelers of TV's *Star Trek* voyaged to the far reaches of the galaxy in the Starship *Enterprise*. This model of the *Enterprise* was donated to the Museum by Paramount Studios. For now, starships exist only in flights of imagination.

(LEFT) **THE HUBBLE SPACE TELESCOPE**

Carried into Earth orbit on a 1990 Space Shuttle mission, the Hubble Space Telescope has revealed spectacular details, from stars and nebulae within our galaxy to infant galaxies at the farthest reaches of the universe. Hubble is designed to operate for a total of fifteen years. This full-size test model of the telescope is on display in the Museum's Space Hall.

(OPPOSITE) **THE EAGLE NEBULA**

Towering pillars of gas and dust lie at the heart of the Eagle Nebula, photographed by the Hubble Space Telescope in 1995. Amazingly, the height of each pillar is more than a light-year—the distance light travels in a year, or about six *trillion* miles. The Eagle Nebula is located 7,000 light-years from Earth in the constellation Serpens (The Serpent).

journeys require not only new technologies but a deeper understanding of energy and matter. Even the relatively modest plans for the coming years and decades call for a certain optimism. But that is the nature of the frontier. The most important message the National Air and Space Museum can inspire isn't written on the exhibit labels: our achievements can be as limitless as the sky itself. The same kind of genius that invented the airplane will one day create the starship; perhaps its inventor will be one of the countless people who come to the Museum and stare, in wonder, at the fruits of human ingenuity. The dream of flight has lost none of its power by coming true. It is the reason why the sight of an airplane passing overhead can still seem like a miracle. ■

PHOTOGRAPHY CREDITS

Photographs are listed by photographer, when known, then by page number. When available, Smithsonian Institution negative numbers are given in parentheses.

The following photographs are courtesy Smithsonian Institution: v; vi; vii (95-8270); viii; 4T; 4–5B (89-13685); 5T ; 6R (87-6620); 7TR (89-11925); 7BL (A-30915-H); 8B (97-15315); 9T; 9B; 9BK (16786); 10 (A-43268); 10BK and 11(85-11473); 12TL (A 28131); 12BR (97-15316); 12BK (A18110A); 14 (painting by Garnet W. Jex. 42969/96-15732); 15 (Science Museum, London, A-45975); 16TL; 17 (USAF, Wm. J. Hammer Scientific Collection, A-4189); 18 (86-3018); 19 (A-43395A); 20L (A41899-B); 20TR (A-2708G); 22 (A-18824); 23T (A18801); 25 (A26767B); 26T (97-15333); 26B (USAAC, A-42296); 26BK (97-15332); 28T (A-38650E); 28B (A41703); 29 (A-317B); 30 (88-16089); 31 (96-16069); 32 (75-7730); 32I (72-11123); 34 (80-2385); 35 (A-42667-E); 36 (86-13505); 36I (A-31980); 37 (97-15334); 39T (80-12297); 39B (80-12296); 40 (89-21553); 41T (79-4639); 41BL (91-17304); 41BK (A-45178); 42 (96-16167); 43 (78-11954); 44 (87-10389); 45TR (96-16073); 45BL (96-16172); 46T (A-42906-A); 46B (A-3491); 47BL (97-15300); 47CR (source: NASM 326 Cat. 1935–44. Framed original currently hanging in NASM Gallery 208 "Pioneers of Flight." 89-21352); 48TL (92-3139); 48TR (96-16171); 48B (A-38470-D); 49BL (89-22371); 49BR (90-2378); 49BK (A31281D); 50TL (85-17117); 50B (90-15729); 53L (80-15324); 53R (A-33652-E); 54 (A45760); 55 (A-32036M); 56T (82-1329); 56B (A-33811D); 57TR (93-4226); 57B (96-16166); 58 (United Technologies Corp., 90-2133); 59 (*Breakthrough Over Kiev,* by James Dietz); 60 (1915 Farré, 90-9745); 61 (postcard view: *Alpine Landscape with Soldiers and Biplane,* by E. E. Walton in the NASM Art Collection, 66520/90-327); 62T (90-4832); 62B (80-3101); 64R (85-

12318); 65L (87-15489); 65C (90-8271); 65R (94-7859); 66L (91-17361); 67L (A-3853); 67R (76-13317); 68L (credit: Edmond Petit, Icare); 68TR (91-3052); 68BR (96-16296); 69 (97-15314); 71B (97-15280); 71R (A-51324); 72T (90-9734); 73; 74TL (91-3490); 74TR and BK(96-16332); 74B (96-16338); 75B (91-3488); 76T (NASM Public Affairs Office); 76B (96-16295); 77 (96-16336); 79T (Peter M. Grosz, 97-15282); 79B (Peter M. Grosz, 97-15278); 80B (Peter M. Grosz, 97-15279); 81T (81-16894); 82T (96-16335); 82B (96-16337); 83TL (77-54); 83BK (A-1596); 84TL (Peter M. Grosz, 97-15283); 85TL ; 85BR (82-11868); 87 (A-336); 88I (A-12720); 89T (A-12720); 89BL (97-15049); 89BR (78-7507); 91L (82-3646); 91R; 92TL (N. Dewell Collection, 75-7024); 92BK (75-15168); 92BK (USAF Photo Collection, 2040 AC); 93B (76-2796); 94 (97-15335); 95TL; 95BR; 96BR (43532); 97BR (National Aeronautics Association); 98 (97-15338); 99T (Sherman Fairchild Collection, 91-14177); 99B (80-2104); 103TL (95-8719); 103R; 103BL (97-15365); 104B (80-3182); 105 (89-5925); 106 (85-19406); 107L; 107R (97-15047); 108T (90-6693); 108B; 110 (97-1586); 111L (88-12579); 112; 113T (95-3544); 115 (83-15530); 116 (97-15046); 118R (85-13091); 118BL; 119TL (78-5520); 120L (96-16068); 120TR (89-5203); 121 (97-15048); 123TL (78-13934); 123B (96-15162); 124 (reprinted by permission of The New York Times Company, 89-21559); 125 (Rudy Arnold Photo Collection. Rudy Arnold negative no. 5205/NASM acc. no. XXXX-0356); 127 (USAF Photo Collection, 52927AC); 128 (USAF Photo Collection, 67022AC); 129R (USAF Photo Collection, 25157 AC); 129BL (USAF Photo Collection, K2407); 130TL (88-10667); 130BR (85-18328); 135 (USAF Photo Collection, K241); 136 (USAF Photo Collection, 37212 AC); 138TL (81-896); 138B (USAF Photo Collection, 50824 AC); 139; 140T (USAF Photo Collection, 30575 AC); 140B (USAF Photo Collection, 160449); 141T (95-1660); 141BL; 141BC; 141BR; 143TL (USAF Photo Collection, K1304); 143BR (Boeing Aircraft Co., photo no. P4820, via

C. G. B Sturat Collection, a-868); 144 (95-9001); 145–46 (92-6535); 147 (USAF Photo Collection, 54919 AC); 148 (USAF Photo Collection, 72989 AC); 148I (USAF Photo Collection, K3909); 149 (97-15309); 150T (86-2149); 152T (USAF Photo Collection, 148446 AC); 152B (USAF Photo Collection, 53690 AC); 153T (USAF Photo Collection, K2482); 154 (90-6597); 155I (84-10037); 156T (USAF Photo Collection, 54530 AC); 156BL (USAF Photo Collection, 53656 AC); 156BR (USAF Photo Collection, K2170); 158T (97-15307); 159 (85-7301); 160L (97-15281); 160R (97-15174); 163T (97-15175); 163B (85-16191-19); 164 (97-15172); 166 (90-5880-32); 167 (75-8859); 168 (74-11297); 170T (USAF Photo Collection, 36354 AC); 170CR (USAF Photo Collection, 55305 AC); 171BL; 172 (USAF Photo Collection, 57193 AC); 173B; 174; 175 (76-7559); 176L (94-8270); 176R (97-15292); 178 (97-15287); 179L (79-6559); 180 (USAF Photo Collection, K6522); 181 (USAF Photo Collection, K5027); 181I (USAF Photo Collection, 34504 AC); 182I (USAF Photo Collection, 39604 AC); 183 (USAF Photo Collection, K5138); 184 (USAF Photo Collection, K7050); 186T (77-12796); 186B; 187 (Frank Winter Collection. A4075D); 189 (97-15291); 190 (97-15308); 191T (Richard Rash Collection. 97-1357); 192 (88-87); 193T (Boeing Aircraft Co., 97-15276); 193B (97-15310); 194BK (92-1303); 197 (Frank Winter Collection. Copyright © 1957 by The New York Times Company. Reprinted by permission.74-11620); 199 (97-14645); 199I (Reg Keniston photo. Frank Winter Collection. 97-15317); 200L (Frank Winter Collection. 83-309); 200TR; 202 (96-16049); 203 (97-15288); 203BK (97-15290); 204 (97-15277); 205TL (Dwayne Day); 205BL (97-15289); 206 (48826-A); 212TL (74-12209); 212 (A3759); 215 (85-17539); 217T; 219B; 239 (79-765); 244BR (80-4978); 246 (IMAX Corp., © 1994 Smithsonian/Lockheed Corporation); 251; 255-6 (97-16075); 259; 262; 263B; 264 (85-17427); 274TL; 274R; 275B; 276B (97-15343); 277B (97-15344); 279 (NASM Public Affairs Office); 282T (96-15199); 282B; 285 (94-4522); 297; 297IR; 297IL;

299TR; 299L (93-2834); 303I (93-2813); 304L; 304TR (92-7701)

Mark Avino (SI-OPPS): x (1967-0176 NASM 1794); xii; xiii; xiv; 2; 6L (95-8900); 13; 52 (88-8621); 67BR; 78; 80T; 81B (86-12092); 86; 88; 97TL; 103BL (97-15365); 117; 122TL; 131 (97-15368); 151 (97-15363); 161; 173T; 182; 228; 244TL; 244TR (97-15171-4); 253BR; 280B; 281B (97-15366, 97-15340); 283B

Ross Chapple: 47T (SB1926); 113B; 114; 119B; 120BR; 150B; 171T (from *The Smithsonian: 150 Years of Adventure, Discovery, and Wonder,* by James Conaway. Smithsonian Books © 1995 Smithsonian Institution): 242L

Richard B. Farrar (SI-OPPS): 155T (74-4295)

Dale E. Hrabak (SI-OPPS): 49T (79-13577); 75T; 137T (80-17164); 170BL (79-4623); 194L (80-18640); 303

Eric Long: ii; xi; 51 (97-1412); 63; 101T (97-15873); 131 (97-15368); 133 (97-15874); 151 (97-15363); 157 (97-15875); 169B (97-16097); 179R (97-16096); 188 (97-16107); 205TR (97-15881); 214IT (96-502); 236 (Collection of Ross Perot, 97-15886/2); 260 (97-15880); 269I; 277T (97-15872); 281T (97-15366)

Terry McCrea (SI-OPPS): 273 (94-8296)

Dane Penland (SI-OPPS): 93T (79-831); 96T (80-2080); 101B (80-2092); 102 (80-2083); 104T (80-2082); 122B (80-4969); 137B (80-4973); 153B (80-2088); 158B (80-2093); 162 (80-4971); 209 (79-833); 278T (80-13716)

Carolyn Russo (NASM): 71TL; 84B; 111R; 191B; 278B

Evan Sheppard: 70; 83BR

The following photographs are courtesy National Aeronautics and Space Administration: 178 (75-11491); 201 (P-8699B); 210 (S65-34635); 213 (S62-08774); 214L (361-1927); 214IB (MR 3-39); 216L (S62-07654); 216R (S62-304); 219T (JPL-2); 220 (858/S65-46638); 221 (1025/S65-63197); 223T (S66-25779); 223B (JPL-2); 224 (S66-50735); 225 (S66-62926); 226R (1650/S67-19770); 226L (S67-21294); 227 (1669/S67-50903); 230T; 230T; 230T; 230B (S68-50713); 231 (1880/S69-35097); 232 (AS8-14-2383); 233L; 233R (69-H-10/69-HC-12); 237 (JPL-2/AST-01-056); 238 (AS9-20-3064); 240R (2135/S69-31048); 240L (S69-39525); 241 (AS-11-40-5908); 242TR (AS11-44-6576); 243BL (AS11-37-5528); 243TR (AS11-40-5949); 245T; 245T; 245T; 245B (AS11-40-5899); 248 (P10623B); 249TL (AS13-59-8500); 249BR; 252 (AS17-148-22727); 253BL (AS17-140-21386); 253T; 253T; 253T; 254 (SL4-143-4706); 260T (S75-29432); 263T

(JPL-6-15/P-18641); 265 (S81-30498); 265 (S87-26113); 266B (S83-35783); 267 (S84-27562); 268 (5L(S)030 NASA/86-HC-51; 269 (6265/S86-38989/108-ksc-386c-1110/83); 271BR (STS063-711-080); 271BL (S76E5199); 272 (STS071(S)072/95-HC-680); 275T (EC94-42883-4); 283T; 284; 286 (S95-02988); 288-89; 291B; 293T; 293B; 294 (L-91-16, 132/3, photo by Gregory L. Mekkes and Carol Petrachenko); 295 (PR2145/ EC96-43631-2); 298 (1996-P-46480); 300 (P39225 MGN81); 301 (P-21082); 301B (P-48526); 302T (JPL-19-3/Cassini/Huygens/9-43538); 305 (HST20-7/STScl-PRC95-44a). Photos on p. 297 are courtesy of NASA/Jet Propulsion Laboratories.

Other sources:
1 (© Caroline Sheen); 8T (courtesy Museé de l'Air); 57TL 1904, by George Evans and Ren Shields (Bella C. Landaur Collection of aeronautical sheet music, Smithsonian Institution Libraries); 66C (M. O'Connor collection, via J. W. Herris); 66R (collection of Alan Durkota); 90 (© Ted Wilbur); 132 (Carrol [Cal] Stewart); 142 (Quentin Aanenson); 169 (Edwards Air Force Base History Office); 177 (Charles A. Lindbergh Picture Collection, Yale University); 194R (courtesy of Mrs. William Bollay); 196, 198 (both), 211, 217B, 218, 261, 266T, 271TR (Space Commerce Corp.); 208 (Boeing Historical Archives); 212TR (Roger Ressmeyer © Corbis); 234 (RSC Energia, Korolev, Russia); 235 (both photos by David S. F. Portree); 235BK (All rights reserved. © 1997 C. P. Vick); 257BR (courtesy of The Institute of

Space and Astronautical Science); 257L (courtesy of the European Space Agency); 258 (courtesy of *Aviation Week & Space Technology*); 276T, 281B (courtesy of the U.S. Air Force); 280T (Air France); 287 (Lockheed Martin Corp.); 288T (courtesy of Boeing Commercial Airplane Group, K58528); 288B (F-22 team photo by John Rossino); 291T (advanced flying automobile concept by Dr. Branko Sarh); 302B (painting courtesy of the European Space Agency, P-42357 BC)

National Archives: 64L (165-ww-424-p744); 72B (165-GK-909); 126 (80-G-16871)

Library of Congress: 16BR (LC USZ62-66295); 27 (LC USZ62-65459)